Gentlemen's Disagreement

PETER HEGARTY

Gentlemen's Disagreement

Alfred Kinsey, Lewis
Terman, and the Sexual
Politics of Smart Men

THE UNIVERSITY OF CHICAGO PRESS

CHICAGO AND LONDON

Peter Hegarty is Reader and Head of the School of Psychology at the University of Surrey.

The University of Chicago Press, Chicago 60637
The University of Chicago Press, Ltd., London
© 2013 by The University of Chicago
All rights reserved. Published 2013.
Printed in the United States of America

22 21 20 19 18 17 16 15 14 13   1 2 3 4 5

ISBN-13: 978-0-226-02444-8   (cloth)
ISBN-13: 978-0-226-02458-5   (paper)
ISBN-13: 978-0-226-02461-5   (e-book)

Library of Congress Cataloging-in-Publication Data

Hegarty, Peter (Psychology professor)
    Gentlemen's disagreement : Alfred Kinsey, Lewis Terman, and
the sexual politics of smart men / Peter Hegarty.
        pages.   cm
    Includes bibliographical references and index.
        ISBN 978-0-226-02444-8 (hardcover : alk. paper)—
    ISBN 978-0-226-02458-5 (pbk. : alk. paper) — ISBN 978-0-226-
    02461-5 (e-book)   1. Men—Sexual behavior.   2. Men—
    Intelligence levels.   3. Kinsey, Alfred C. (Alfred Charles),
    1894–1956.   4. Terman, Lewis M. (Lewis Madison), 1877–
    1956.   I. Title.
    HQ28.H443 2013
    155.3'32—dc23

                                    2012039521

Blasphemy has always seemed to require taking things very seriously. I know no better stance to adopt from within the secular-religious, evangelical traditions of United States politics, including the politics of socialist feminism. Blasphemy protects one from the moral majority within, while still insisting on the need for community.

Donna Haraway

Contents

Acknowledgments

This book would not have been possible without the imagination, encouragement, and insight of many people. For example, Doug Mitchell imagined a version of the project years before I could even articulate it. The British Academy supported initial archival work in 2003 at Stanford and Indiana Universities, and Simon Ford, Katie Cumiskey, and Robin Garber all provided warm hospitality as I went. Linda Anderson made possible a later trip to the Yale archives. I'm particularly grateful for the help of several archivists, notably Daniel Hartwig at Stanford, Lizette Ann Royer at Akron, Rebecca Hatcher at Yale, and Shawn Wilson at the Kinsey Institute.

Two previous Heads of the School of Psychology at Surrey deserve particular thanks; Jennifer Brown approved a term of sabbatical leave in 2007 that gave me the space and time to dig deep into this work, and Annette Sterr put up with my continued efforts and distracted attention as I acted as her Deputy Head from 2008 to 2011. Abigail Stewart invited me to Michigan in 2006 for a pivotal visit that allowed me to explore my initial ideas with both psychology and women's studies students. Don Brown, Bill McCeachy, and Mel Manis shared valuable memories of their own lives in the history of psychology. Their memories of psychologists E. Lowell Kelly, Lewis Terman, and Dana Bramel particularly informed my thinking.

At various times and places, scholars who know more about diverse areas of scholarship than I do have read drafts of some of these chapters, provided me with opportunities to talk about this work, offered their reflections and criticisms of it, and pointed me down new unexpected directions. Often their intellectual gifts were not obvious to me at the times of our conversation, but I owe particular debts to Stephen Algelides, Naomi Andre, Gavi Ansara, Meg Barker, Peter Baxter, Betty Bayer, Toni Brennan, Karl Bryant,

Geoff Bunn, Jim Capshew, John Carson, Angela Cassidy, George Chauncey, Martin Cheek, Fran Cherry, Victoria Clarke, Liz Cole, Alan Collins, Kate Crehan, Adrian Coyle, Nicola Curtin, Lisa Downing, Michelle Fine, David Gibson, Eoin Gibson, Phil Goff, Gareth Hagger-Johnson, Peter Hair, David Halperin, Ellen Herman, Ian Hodges, Nadine Hubbs, Darren Langdridge, Larry LaFountain-Stokes, Kinsten Leng, Lih-Mei Liao, Ram Mahalingam, Sarah McClelland, Joanne Meyerowitz, Lisa Moore, Julie Perks, Felicia Pratto, Pete Lamont, Michael O' Rourke, Liz Peel, Mike Pettit, Wade Pickren, Hans Pols, Freyja Quick, Graham Richards, Katrina Roen, Joe Rollins, Alex Rutherford, David Serlin, Stephanie Shields, Charlotte Sleigh, Jennifer Terry, Naoko Wake, David Winter, and Leila Zenderland.

Finally, my thanks go out to the team at University of Chicago Press for helping me to bring this project to fruition, particularly Richard Allen for his careful editing, and Tim McGovern for managing the process of publication.

This book is for my mother Nuala and my father Seamus; for my five siblings—the teammates my parents gave me in this life; and for Andrew Penman, the partner I chose for myself. None of these people would expect me to dedicate this book to them, but they believed in its completion when I doubted it the most myself. More than my words, they prove my argument that a bright future is easiest to imagine when life is approached with a team size that is bigger than one.

Materials from published and unpublished papers of Lewis Terman and from papers of Alfred Kinsey (including unpublished correspondence) have been used with permission from Stanford University Press and the Kinsey Institute for Research in Sex, Gender, and Reproduction at Indiana University, respectively.

1

A Gentlemen's Disagreement?

Just what facts would have been convincing about the superiority of the method of ordinary interviewing are not clear to us, and therefore the bald statement that Terman's "Data would have been more reliable if they had been obtained by direct interviewing" (Kinsey, Pomeroy, and Martin, p. 31) is all the more galling. It is proper for some toes to be trodden on in a scientific investigation, but a smile in return for a scientific treading is to be hoped for only when evidence is presented. Authority contradicting authority gets us nowhere.

Cochran, Mosteller, and Tukey, *Statistical Problems of the Kinsey Report on Sexual Behavior in the Human Male* (1954, 76)

To my knowledge, Alfred C. Kinsey met Lewis M. Terman face to face only once, on the campus of Stanford University in 1952. It occurred after a talk that Kinsey had given about his sex research. The meeting was brief, perhaps because these two had said so much to each by other means four years earlier. On January 5, 1948, Kinsey had published the best-selling popular science book of a generation, *Sexual Behavior in the Human Male (SBHM)*. The reported sexual histories of the 5,300 White men within its 804 pages stayed on the bestseller list for months. At $6.50, *SBHM* transformed booksellers' imaginations of just how avidly the public might consume science. *Sexual Behavior in the Human Female*, an equally profitable and voluminous book on the sexuality of women, would follow five years later.

Kinsey's sex research is widely interpreted as a moment in what Paul Robinson (1977) called "the modernization of sex": the shift from religious to psycho-medical authority as to how modern sex lives should be conducted. Like Freud, Kinsey is the kind of author whose writings about sex can appear jarring, arresting, or refreshing just as much today as when they were first written. For example, by using the orgasm (or "outlet" as he would term it) as the unit of sexual behavior, Kinsey found that only 87 percent of middle-aged married men's sexual behavior occurred during marital sex; extramarital affairs and masturbation accounted for much of the rest (*SBHM*, 281). Male homosexuality had been considered a psychopathology that compromised military strength during World War II, and homosexuality among women and men would be codified as a mental illness by the American Psychiatric Association in 1952. But Kinsey reported that a full 37 percent of American men had had at least one homosexual experience leading to orgasm since their adolescence.[1] In the same year that the UN Declaration of Human Rights stated that "the equal and inalienable rights of all members of the human family is the foundation of freedom, justice and peace in the world," Kinsey reported that American farm boys' sexualities were ill-captured by conventional species boundaries. Almost 50 percent had had sex, at least once, with an animal.[2] Generations of scholars have commented upon, used, and critiqued the Kinsey studies. In the first decade of the twenty-first century, Kinsey's life and work became the subject of American documentary, feature film, and fictional novel—each of which contributed to making the network of bisexually active swinging researchers around Kinsey in the small Midwestern town of Bloomington, Indiana into something of a modern American myth.[3]

The other party to this meeting—psychologist Lewis M. Terman— celebrated his seventy-third birthday just twelve days after the publication of *SBHM* in 1948. Terman had retired six years earlier from a long career as chair of the Stanford psychology department. When Terman read Kinsey's book, he was less convinced than the American public, or its press, that Kinsey had achieved a work of scientific genius. Terman worked diligently on a long review of the book, which he submitted unsolicited in May 1948 to Lyle Lanier, the editor of the journal *Psychological Bulletin*. This leading journal had already published a short positive review of *SBHM* by the psychologist Carney Landis (1948). Landis had originally submitted a much longer review which he had shortened at Lanier's insistence. Presented with Terman's lengthy critical review, Lanier now feared his editorship being perceived as partisan against Kinsey if he allowed its publication.[4] Terman revised the review by shortening it, and made it *appear* shorter by presenting a long quotation from Quinn McNemar in smaller type.[5] Lanier accepted Terman's review but reminded Terman that if Landis were to request it, he would allow him to publish his original positive review of Kinsey. To my knowledge, Landis never made this request.

Terman's work was as world-shaping and controversial as Kinsey's.[6] When published in 1916, Terman's "Stanford-Binet" test of children's intelligence became the most successful of several such English-language tests competing for dominance in the United States. Shortly thereafter Terman worked as part of a team of psychologists, led by psychologist Robert M. Yerkes, who tested the IQ of American soldiers en masse during World War I.[7] That army research became a matter of national public discourse in the 1920s, leading Terman to defend IQ tests against criticism in the popular press that the tests were anti-egalitarian and antidemocratic.[8] During the 1920s, as funding for the study of child behavior was more firmly nationalized,[9] Terman developed several group tests that could be used to quickly measure the IQ of vast numbers of American schoolchildren,[10] and he initiated a longitudinal study of high IQ children of unprecedented scope and duration.[11] This study of "gifted children" was a lifelong passion of Terman's, and the data he gathered on gifted children continue to be both mined by research psychologists and debated by historians of psychology today.[12]

As psychologists began to consider how differences in IQ scores both between individuals and between social groups could result from environmental influences, Terman and his colleagues at Stanford came to define the conservative position that individual and racial differences in IQ scores were stable consequences of biological inheritance.[13] Such longstanding debates about IQ are anything but an "academic" issue. In the United States, a person's IQ score can be a matter of life or death; the death penalty cannot be applied, in many states, to prisoners with an IQ lower than 70.[14] However much IQ shapes, saves, and ends American lives, American psychologists lack consensus on the definition of "intelligence," beyond the opinion expressed by Terman's friend E. G. Boring in the early 1920s that "intelligence" is simply whatever intelligence tests measure.[15]

In short, Terman's life and work were not free of controversy, raising the question of why—as an elder statesman in the discipline of psychology—he described his review of Kinsey's book as "without exception, the most difficult I have ever attempted."[16] Terman's review presents a puzzle to the history of sexuality, intelligence, and psychology. One of Kinsey's biographers has called Terman Kinsey's "most determined critic,"[17] and while the exchange between these two is mentioned in biographies of each of them, the *particular* matters that arose between them have not been attended to in the literatures about either.[18] The most detailed published history of Terman's review of Kinsey appears in James Jones's biography *Alfred Kinsey: A Public/ Private Life*. Jones described the review as "vigorous, but gentlemanly" (1997, 589) and Terman as a reviewer who was "aware that he had strong feelings about the book" but who "struggled to keep them from creeping into his writing, a challenge that required several drafts" (588).

However, there is more than a hint of ungentlemanly politics about the review and its aftermath. While Terman is often remembered as a founding father of the IQ testing movement, he also became a dedicated sex researcher in the years between the World Wars, publishing books on the measurement of gendered personalities and marital happiness. Kinsey had reviewed Terman's research very negatively in *SBHM*. (In contrast, he had praised psychologist Carney Landis's research on the sexuality of disabled women.)[19] Terman's private letters suggest that strong feelings increasingly determined his actions once he learned that his criticisms of Kinsey were shared by others. Terman studied as an undergraduate at Indiana University from 1901–3, where Kinsey worked throughout his academic career. In 1949, Terman brokered his alumnus status to criticize Kinsey to the IU president, Herman Wells.[20] Terman also wrote to psychologist Robert Yerkes whose opinion of Kinsey mattered particularly.[21] Yerkes and Terman were close friends and frequent correspondents from World War I until Yerkes' death in 1956. After World War I, Yerkes chaired the National Research Council's Committee for Research for Problems of Sex (the CRPS).[22] Throughout the 1920s and 1930s, this committee had strenuously avoided controversy by refusing to use Rockefeller Foundation funding to support the research of social hygienists who aimed to conduct what we now call "sex surveys," funding hormonal studies of sex and studies of animal behavior instead.[23] Terman's professional relationship with Yerkes allowed him to become an exception to this rule, and he garnered CRPS funding for research on "masculine" and "feminine" personalities, marital happiness, and the adult sex lives of gifted children.

Yerkes retired from the CRPS in 1947, passing the chairmanship of the committee to George Corner. When matters threatened to become ungentlemanly between Kinsey and Terman, the legitimacy of the CRPS came into question once again. The unparalleled popular success of Kinsey's *SBHM* had made it obvious that there had been a large national investment in the sex survey. However, the critical review by the venerable psychologist Lewis Terman, in whom the CRPS had also invested, cast doubt on the legitimacy of such research and the legitimacy of the CRPS's definition of "good science."

After reading Terman's review of Kinsey, Yerkes wrote to Corner about the need to "safeguard values in accordance with accepted scientific practice,"[24] but this matter was not simple. The Terman/Kinsey debate made clear that the two scientists who had been trusted with CRPS grant money could not agree on which scientific values to safeguard and which to question when empirical science was extended to human sexuality.

Why had the CRPS come to invest so heavily in Kinsey in the first place? Robert Yerkes first met Kinsey at a meeting of the American Association for the Advancement of Science in 1941, at a time when the kinds of biomedical research projects that the CRPS had been funding were being supported differently and the raison d'être of the CRPS was in question.[25] By 1941, Yerkes needed to invest in a different kind of sex researcher. A few years earlier, Kinsey—an insect biologist by training—had taken his first steps into sex research by teaching a "marriage education" course to his IU students, and had begun to gather his students' "sexual histories," initially with questionnaires and later with one-on-one interviews. Kinsey had presented a paper at the AAAS on diverse sexual histories of men that he recruited on campus, in Indiana's penitentiaries, and in Chicago's gay bars. Kinsey had approached the task of collecting sex histories with an enthusiasm for gathering large samples through tireless fieldwork that he had previously applied only to his insect biology work. His AAAS paper (1941) showed variability in men's histories that would bedevil any attempt to divide men into two clear species of "homosexuals" and "heterosexuals." To Yerkes, Kinsey seemed to be the right scientist for the job of keeping the CRPS in play. Through the war years and after, the relationship between Bloomington and the Rockefeller Foundation thickened. Kinsey's team expanded, and his research came to consume half of the CRPS's annual research funding budget.[26]

Quite apart from their explicit disagreements, Kinsey and Terman were similar in many respects. Like so many other American scientists, both were caught up in the enthusiasm for eugenics in the 1920s. This social movement drew on Darwinian theory in its aim to increase national, racial, or species "fitness," found allies among the IQ testing movement, and led to sterilization laws in many American states.[27] Indeed, both the biology department where Kinsey worked at IU and the psychology department at Stanford that Terman chaired from 1922 until his retirement in 1942 were housed in buildings named Jordan Hall to honor the biologist and eugenicist David Starr Jordan, who had been president of both universities. Curiously, Kinsey— the biologist—adopted quite a behavioral approach to male sexuality, while Terman, the psychologist, argued strongly that intelligence was a matter of biological inheritance. Both suffered from childhood illnesses that have been interpreted as influential on their mature scientific work, including their beliefs about the extent to which biology provides a blueprint for individual destiny. These two men also crossed trails in research; both conducted research on sex offenders in San Quentin, California's oldest prison.

Beyond the specifics of their intersecting lives, Terman's debate with Kinsey has intrigued me because it points to the unstable ways that two ineffable, but often governed, attributes of people, "sexuality" and "intelligence," came to signify each other in the human sciences of the early twentieth century. Both "sexuality" and "intelligence" exemplify the paradox identified by historian Steven Shapin that all attempts to separate "sciences" from "politics" engender sciences that are effective politically to the extent that they are taken to be nonpolitical.[28] My point in drawing Kinsey and Terman together in this book is to make the critical histories of these two constructs speak to each other more explicitly. Indeed, I will argue that silences about the ways intelligence and sexuality have been co-constructed have been constitutive of both concepts. But to avoid introducing a silence over my own stakes in this debate, let me explain how I came to be interested in these two and to write this book.

In the 1980s and early 1990s, prior to an economic boom-and-bust in the Irish knowledge economy, public conversations about Irish intelligence were organized by fears of a brain drain caused by emigration.[29] In 1993, I contributed to brain drain statistics when I took my Irish undergraduate degree in psychology to California to begin a Ph.D. program in the Stanford psychology department that Terman had once chaired. I came out about six months after moving to Stanford, becoming another ordinary person for whom life narratives of migration and the realization of being gay are connected.[30] Being both gay and a foreign student fueled my developing sense of being "other" than "Stanford material," a phrase used freely, and only sometimes with ironic detachment, to describe those who were read as displaying the intellectual attributes that justified their presence. Much like the feminist psychologist Sandra Bem (2001), who was a faculty member there in the 1970s, I experienced the Stanford psychology department as a place where there were only two grades, A+ and F. But Stanford was also an environment where many more bright people needlessly feared the F than their peers might have guessed. People who knew me during my time at Stanford might remember someone who was more often alienated from the institution's norms than anything else.

The early 1990s were an interesting moment to arrive at an American campus as a foreign student, with chest x-rays in hand—the visual evidence that I had brought no infectious diseases with me. This was the era in which terms like "political correctness" gained purchase as a means of silencing debates on campus about inequality about sexuality, gender, and race.[31] My prior knowledge of such inequalities within the United States left me ill equipped for such conversations; one of my first "consciousness-raising" moments after coming out was a tense conversation in which I (now to my shame) argued that a college course that considered queer theory was obviously a case of "political correctness" taken too far. My interests changed in this climate, as I increasingly came to perceive psychology's meta-theoretical commitment to cognitivism as being in tension with the exciting ideas about modern thought that I was absorbing from queer theory and feminist philosophy.[32] From Ireland, I had originally proposed Ph.D. research that would examine how schoolchildren would estimate their capacity to solve math problems. I did not lose interest in studies of people's thinking, or in their importance for growing children. But I ultimately wrote a Ph.D. about the ways that heteronormativity shaped the interpretation of developmental research about sexual identity formation.[33]

I spent a lot of time in the 1990s trying to figure out whether running experiments on Stanford undergraduates was a smart thing to do, and trying to think through the politics of conducting behavioral experiments on people more generally. As a result of these productive quandaries, I was often to be found looking for a different story about psychology in the extensive Stanford library instead, lavish even after the effects of the 1989 Loma Prieta earthquake. There, I read the history of psychology and the emerging poststructuralist critique of the discipline by discourse analysts, rhetorical psychologists, and others with enthusiasm and earnest appreciation.[34] Since George Stockings' early warning (1965) about the way that intellectual history can reify contemporary states of understanding in the human sciences, American historians of psychology had moved away from a model of the discipline's past as an accumulation of facts about the timeless individual psychological subject. Thomas Kuhn's notion of paradigm shift as the mode by which scientific consensus changed had led some historians to think of psychology as a nonaccumulative science.[35] By the 1990s, past psychologists' theories, including Terman's, were increasingly described as "locked in time,"[36] consistent with social constructionist challenges to the natural science model of psychology.[37] New work in the history of experimental and statistical methods both threatened some of the seemingly "hardest" elements of experimental psychology and opened up new lines of inquiry.[38]

Along with the poststructuralists, some historians began to recognize how they themselves *produced* rather than documented the past, and how historical accounts of psychology in the here and now were interventions in debates in the present about how psychologists might best imagine and analogize human subjectivity and action.[39] The history of psychology, such as that which Terman's friend E. G. Boring had taught and wrote on the subject at Stanford,[40] no longer aimed at celebrating great men of the past. It had become a less "internalist" endeavor—and instead a project that grasped at complex epistemological and ontological questions about the relationship between psychology and history.

The more I read, the stranger the practice of experimenting on undergraduates appeared. Back at the Californian Jordan Hall, I also began to see how the experiments that I was conducting could be interpreted as *performative* in a way that jarred with the positivist narrative of scientific progress which organized psychology at Stanford.[41] The embodied day-to-day double life of a reluctant experimentalist and dilettante historian helped me to realize that historians of psychology were not alone in their interest in the contested zone of psychology's past.[42] The narrative form of the psychology journal article, a genre I was slowly learning to write, required the writing of a modernist story in which past work is preparatory for the new findings, and the conclusions must always include the future-focused message that "more research is needed."[43] Those psychologists who might be thought to be most resistant to historicism—the lab-locked experimentalists—have always been busy writing self-serving "internalist" narratives in their myriad reporting of new empirical results.[44] All of these reflections on my own scientific practice led me to be concerned with differences *within* scientific psychology in the knowledge, narratives, and power that might be brought into being by our work.

Unexpectedly, my attempts to square the demands of Stanford's experimental psychology Ph.D. program with my changing perspective on psychology's history led me to an interest in Lewis Terman, who had first brought that same department to prominence in the 1920s. In my own experiments, I was testing a hypothesis that the ways that heterosexual undergraduates might talk about their childhood gender experiences would be different when those students were placed in conditions where they might be misperceived to be lesbian or gay. These experiments were an attempt to respond to the rise of neo-essentialist theories linking adult sexual orientation with childhood gender experiences, and to bring some ideas about the performance of gender into conversation with experimental psychology.[45] My search for origins of these neo-essentialist theories led me back to Anne Constantinople's important critique (1973) of the conflation of gender-typing with gender identity. Through Constantinople's work, I was led to a seemingly foundational attempt to measure "masculinity-femininity" by asking people about their behaviors and interests: Lewis Terman and Catharine Cox Miles's 1936 volume *Sex and Personality*.

Toward the end of my studies at Stanford, I wrote a chapter that attempted to update earlier feminist criticism of that book in light of the explosion of work in lesbian, gay, and queer studies which characterized the period.[46] To engage in lesbian, gay, and queer work at Stanford in the 1990s was to debate the works of Michel Foucault, one of several historians who, in response to the sexual revolution, described how sexuality had undergone a process of secularization and become core to the modern, Western, industrialized self.[47] Foucault introduced a way of thinking about sexual politics as organized by a dynamic of the production of expert discourses about sex rather than the repression of natural sexualities. In Stanford in the early 1990s, Foucault was read avidly and expertly by groups of graduate students in the departments of history, English, modern thought, and anthropology. However, the disciplinary culture in the department of psychology was very, very different. Within the English-speaking world, British historians of psychology and critical psychologists had been quicker to recognize the importance of Foucault's approach to power than their American counterparts.[48] Critical psychologists' attention was particularly captured by Foucault's genealogical study *Discipline and Punish* (1977b), which specified how "disciplinary power" worked anonymously to control people with surveillance that specified the particulars of their conduct and their psyches. Foucault's work had been appealing particularly to early critical psychologists who were concerned to collapse the distinction between "the subjective" and "the social,"[49] and it appeared "dark" later to some historians of psychology who read his work as leaving little space for individual agency and historical change.[50] Ironically, some North American feminists critiqued Foucault for the opposite reason; his writing was individualist and did not pay enough attention to structural power.[51] I came to see critical psychologists' rather "dark" reading of Foucaultian power as limited, because they overlooked important ways that his work leaves room for maneuver within systems of power.[52] Through North American lesbian, gay, and queer studies, I became familiar with a "lighter" Foucault. For example, authors such as David Halperin argued that Foucault had suggested new ways of analyzing and transforming the challenges and affordances of living queerly in the first decade of the HIV/AIDS epidemic.[53]

9

I continue to think that critical psychologists are missing a trick in failing to study the Foucault of lesbian, gay, and queer studies, and I hope this book will provide new uses of Foucault's work for those who long for a more critical form of psychology that is not afraid to acknowledge debts to sexual minority traditions of intellectual thought.

I developed the idea for this specific project in 2001, two years after I left Stanford. At that point, I was teaching on the East Coast of the United States and had the opportunity to return to the Terman archives to look again at the material pertaining to *Sex and Personality* in greater depth. While there, I read Terman's review (1948) of Kinsey's *SBHM* in the Stanford library, and became both vexed and intrigued; I couldn't place its particulars in the context of Terman's larger body of work. In that review, Terman attacked the truth status of Kinsey's statistics, the validity of his methods, and the purity of his scholarly intentions. Kinsey's co-author, Wardell Pomeroy, later described Terman's article as a maddening "yes, but" review (1972, 290), perhaps because Terman began by describing how the book's "merits have been recounted so extensively by others," continued with faint praise for Kinsey's "zeal and perseverance" (1948, 443), and then gave himself over to sixteen pages of unrelenting criticism. Yet Terman came to bury Kinsey, not to praise him, and he measured the grave with the yardstick of psychologists' research standards—standards which Kinsey had explicitly rejected, and for somewhat principled reasons.

Some of Terman's points were held in common with other criticisms of Kinsey's work. Terman charged that Kinsey had standardized (or "operationalized" in the emerging parlance of psychological testing) human sexual behavior as "outlet," but had failed to standardize or record the wording of his interview questions; nobody could ever check if his study was replicable. His sample did not match the US population. His attempts to calculate the sample size needed to reliably estimate the sexual behavior of a group had gone badly awry. His interview method also placed a burden of denial on interviewees by asking them when, rather than if, they had ever engaged in particular sexual behaviors, a method that assumed "that everyone has engaged in every type of activity" (*SBHM*, 53; quoted in Terman 1948, 445). In short, the study was too closely tied to Kinsey himself, and the intelligence garnered from men of dubious moral standing too untrustworthy, to build a scientific project that was free of politics. Kinsey's work fell short of the kind of disembodied, transparent, objective methods in which psychologists had learned to invest their limited trust (like IQ tests, for example).

There was nothing about any of that that particularly vexed me. Criticisms of Kinsey's sampling and interview methods were common enough in his own time and in ours. However, Terman's review had particular features that left me scratching my head. Terman seemed to have an interest in some data patterns in *SBHM* that other critics were happy to leave alone, and he staked some odd claims in the course of challenging Kinsey on these points. In the middle of his review, Terman listed four particular points of contention. First, he critiqued Kinsey's assertion that boys whose sex lives began early—typically through masturbation—had more sex later on and were more robust and healthy than their peers. Second, Terman similarly countered Kinsey's claim that rates of premarital intercourse were stable across the generations. Third, Kinsey had staked a claim that men's early sexual experiences appeared to predict later class status; men who *would become* middle class behaved as middle class men did even in their adolescence. Terman argued that these data patterns were incredible, and that they made Kinsey's entire data set untrustworthy. Finally, Terman suggested that Kinsey was wrong that religion *could not* effectively repress sexual behavior. I try to explain each of these "stubborn particulars" in chapters 2, 3, 5, and 6 below,[54] particulars that were forgotten as statisticians attempted to sort the wheat from the chaff in this debate (as I discuss in chapter 7). This list is all the stranger for Terman's lack of attention to Kinsey's claim that "one male in three of the persons that one may meet as one passes along a city street" had enjoyed homosexual sex to the point of orgasm in adulthood (*SBHM*, 623). This omission is examined in chapter 4. In chapter 8 I pick up the matter of how the unresolved issues between Terman and Kinsey continued to matter even when I came to the United States myself in 1993.

How might *Foucault* help historians of sexuality and historians of psychology to make sense of these "stubborn particulars"? In the course of writing this book, I grew to think of Foucault as an enabling historian because he gave particular reasons to approach historical inquiry with open-ended goals. In one pivotal paper, "Nietzsche, Genealogy, History," Foucault (1977a, 76– 100) followed Nietzsche in eschewing a search for ahistorical origin points as the ground for either ethics or epistemology. Instead, he laughed at the search for such universal truths (79) and focused on shifts and transformations in what is taken to be "truth" that continue to be relevant, but go unnoticed. The goal of genealogy, according to Foucault, was "to identify the accidents, the minute deviations—or conversely, the complete reversals—the errors, the false appraisals, and the faulty calculations that give birth to those things that continue to exist and have value for us" (81).

Historians of sexuality of the last few decades have tended to either follow

CHAPTER ONE

Foucault's genealogical orientation to the relationship between the present and the past, or to have independently found their own reasons for adopting an approach to the past that might also be considered "genealogical."[55] However, critical psychologists who have used Foucault's orientation to *discourse* as the implicit unwritten network of assumptions within which knowledge achieves coherence have tended to overlook the important temporal element of the genealogical orientation.[56] Terman's review of Kinsey seemed to me to contain particular elements that had been overlooked as minute deviations. Might these serve as the starting points for a genealogy? This book argues that these forgotten small points between these two men provide pivotal vantage points from which components of much larger androcentric discourses linking sexuality and the intellect might be glimpsed. By using these pivot points to re-map the history of psychology, my aim was not so much to correct errors in our understanding of the past, but to re-narrate the past to affect the direction of current conversations between critical histories, psychology, and the study of sexuality and the intellect.

In writing about Terman and Kinsey genealogically, I concluded that some concepts which had emerged from Foucault's genealogies themselves required genealogical rethinking if they were going to make useful sense of the *dispositif* of sexuality and intelligence in which Kinsey and Terman found themselves at odds.[57] The first of these follows from my "internalist" view from within psychology of the different forms of power/knowledge that might be at work when people make up "psychological" knowledge about each other. Foucault's book *Discipline and Punish*—his genealogical study of the origins of the prison and modern notions of justice and punishment—was enormously influential in shaping early critical psychologists understanding of how people are subjected to power by being made "subjects" of psychological knowledge. Foucault described a shift from "sovereign power" to a modern regime organized by "disciplinary" knowledge, which rendered powerless people yet more powerless by making them visible targets of power: "As power becomes more anonymous and more functional, those on whom it is exercised tend to be more strongly individualized" (1977b, 193). Foucault had psychological knowledge and practice squarely in his sights in his account of disciplinary power, continuing that "all the sciences, analyses or practices employing the root 'psycho-' have their origins in this historical reversal of the procedures of individualization" (193), and stating that normalization worked on visibly exceptional people particularly. As critical psychologist Derek Hook nicely summarized it, "normalization, simply put, is the ultimate goal and effect of disciplinary technologies" (2003, 611). For the Foucault of *Discipline and Punish*—"visibility is a trap" (1977b, 200).

Foucault repeated the hypothesis about power and visibility-within-the-psy-disciplines in *History of Sexuality, Volume 1*. This book was as canonical as anything else within the emerging field of lesbian, gay, and queer studies in North American in the 1990s. Therein Foucault described sexuality as a "centrifugal" system that radiates outward from the normative married couple to ontologize and regulate those constructed as "other." As a form of disciplinary power, "sexuality" made visible those who did not adhere to its norms. In contrast to Christian codes that focused on the behavior of married couples, the modern notion of "sexuality" spoke most loudly about people and practices that were other to the normative married couple; masturbating children, hysterical women, the out-of-control sexuality of the poor, homosexuality and other perversions. Hence, for Foucault, pushing past taboos and speaking the truth of sexuality in the face of repression is likely to rely on claims about the "nature" of such sexual differences, and to extend disciplinary power rather than to escape it.

In *History of Sexuality, Volume 1*, Foucault established the Catholic confessional as the paradigmatic "positive unconscious" of modern Western science about sex; that part of that science which "eludes the consciousness of the scientist and yet is part of the scientific discourse" (1971, xi). Foucault suggested five procedures by way of a description and explanation of that historical transformation; the clinical codification of sex, the endowment of sex with a causal power to cause just about anything, the positioning of sex as a latent obscure force, the requirement for expert interpretation and expert interpreters, and the medicalization of the relationship between normal and abnormal sex (1978, 65–67). Foucault's work has oriented a generation of historians of sexuality to think about modern Western secularization in terms of "the procedures by which that will to knowledge regarding sex, which characterizes the modern Occident, caused the rituals of confession to function within the norms of scientific regularity" (65).

Kinsey's work is a particularly good test of Foucault's confessional model. Through writing this book, I became less convinced that Kinsey's sex research—so central to American narratives of sex's modernization—fitted into the confessional paradigm. The "darkness" of disciplinary power is mismeasured by failing to notice the difference between the confessional practices that Foucault described and the ways that Kinsey used people's "will to knowledge" about sex to affect the direction of the sexual sciences. Individuals certainly "confessed" matters to Kinsey and his interviewers in the sense that they revealed things in private that they often considered sinful. Moreover, the researchers codified those narratives prolifically. However, these confessions were unaccompanied by promises of individual normalization or redemption. Rather, many of Kinsey's most trenchant critics, such as Lewis Terman, were concerned that this new *scientia sexualis* did not normalize sex enough. Kinsey's work did not lack a positive unconscious. But, following the recent work of historian of psychology Graham Richards (2011), I think it is timely to question some assumptions about secularization implicit within the work of postmodern theorists such as Foucault. Qualifying the very austere image of the confessional, Foucault himself urged historians to think about how secular confessional practices had led to a "production of truth, intimidated though it was by its own scientific model," had "multiplied, intensified, and even created its own intrinsic pleasures" (1978, 71). Clearly, the confessional is not the only religious practice that can organize our understanding of the sexual sciences. Here I propose a less *monotheistic* metaphor for the unconscious of Kinsey's science, one that emerged as I developed my understanding of his debate with Terman. I present it here in the hopes that it might incite questions about psychology and history that differ from the now-familiar Foucaultian ones.

Terman's work on Intelligence also gave me pause to question some Foucaultian assumptions about disciplinary power. In *The History of Sexuality, Volume 1*, modern Western sexuality was described as particular because it singularly united diverse anxieties about the health of individual (middle-class) subjects and the demographics of the (largely working-class) populations, where the body and mind meet, and where speech can represent the "truth" of psychic life which is analogized as "inner" to the person. In twentieth-century America, intelligence was also a point where knowledge and power made each other up, questions about individual merit and the health of populations signified each other, and the mind and the body were often imaged to touch. It is easier to make the argument that in the United States—more than in other nations—that "intelligence" provides a second pivot that rivals "sexuality" for its importance within histories of power.[58] Indeed, several historians have characterized intelligence testing as a paradigmatic Foucaultian disciplinary technology.[59]

To be sure, "Intelligence" could not exist without those forms of normalizing measurement upon which modern psychology chronically depends. However, discourses about "intelligence" predate the measurement of "Intelligence" and they continued to compete with IQ testing for legitimacy since Intelligence Testing became an established practice.[60] Crucially, the observation of "abnormality" in the domain of intelligence does not always incite a normalizing impulse, as Foucault's description of "disciplinary power" suggests. When children or adults are ontologized by psychologists as abnormally or precociously "gifted," the modern impulse has sometimes been to support the flourishing of their abnormality, and sometimes to normalize it out of existence. Kinsey and Terman can show us, I think, why the impulse to normalize *sexuality* makes all the difference to modern responses to the ways that these particular people are treated.

The history of statistics provides a clue as to the intersecting dynamics of normalization. Statistical thinking about "populations" exploded in the early part of the nineteenth century. An avalanche of statistical aggregation in the 1830s about social life aimed to address the ideological dilemma in new democratic nation states in which individuals were "free" but needed to be governed and to live in stable societies all the same. Statistical aggregation allowed both for individual freedom and culpability, and for the prediction, government, and control of stable aggregate *rates* of crime, birth, death, suicide, and prostitution.[61] The Belgian statistician Adolphe Quetelet inferred that events occurred with varying probability in accord with what Karl Pearson later called "normal" distributions, with few examples of very extreme values and most values occurring around the average. Within this understanding of normality, rare deviations from the average were seen as destabilizing effects of individual freedom and threats to the stability of "society."

A second anxiety about instability and population arose in Malthus's *Essay on Population* (1798), which proposed that populations were likely to increase unless checked by historical disasters. Malthus was a key influence on Charles Darwin's theory of evolution, in which natural and sexual selection acted on natural variation in populations to "select" those characteristics of organisms most likely to survive and flourish.[62] Darwin opened up a new discourse about natural variation as the raw material upon the selection of the best forms for the future depended; variation was necessary for populations to evolve.[63] Darwin's cousin Francis Galton spotted the relevance of this view for understanding differences in what we now might call "intelligence." His science of eugenics expanded this thinking about variation back to nation states and populations, with a mix of evolutionary optimism and despair about what might happen if human reproduction was left unregulated to matters of individual freedom. In Galton's eugenic thought, some kinds of extremely unusual people—such as the geniuses and great scientists of the British elite—were naturally different from the rest of us, but were also less likely—by dint of social circumstances—to have children who might inherit their talents. In Galton's view, such people deserved protection and privilege and ought to be obliged to multiply. Galton also extended Quetelet's observations about normal curves from the bodies of soldiers to anthropometric measurements of the minds of the public.[64] He encouraged the American psychologist James McKeen Cattell to develop his ideas that such tests of sensory acuity and speedy reactions to stimuli could be measures of something akin to the modern concept of "intelligence."[65]

I think that Foucault's description of "normalization" in *Discipline and Punish* makes more sense of the former Queteletian socially conservative definition of the "normal" oriented around the average and less to the latter Galtonian definition of the "normal" oriented around ideal cases and the promises of the future. I do not mean that there are no power dynamics involved in Galtonian normativity, but only that there is a good reason to tease apart these two forms of normalization as different. Galtonian normativity is at least a trap with an escape latch in the *positive* description of atypicality. This book could be read as a history of what happened when Terman sprang this latch apropos of precocious intelligence and when Kinsey sprang it again apropos of precocious sexuality. Perhaps Terman wrote his review because his springing of the first latch, in his study of gifted children, depended upon bolting down the second so that precocious sexuality would not become an object of Galtonian normalization.

These intersecting dynamics are also suggested by *History of Sexuality, Volume 1.* In that book's final chapter, Foucault argued that the history of modern sexuality demonstrates that there is no clean historical break between the regimes of sovereign and disciplinary power; he described (1978, 147–49) the Marquis de Sade, eugenics, biological state racism, and the Nazis as evidence that disciplinary power, more commonly adherent to sexuality, has, on occasion, been organized around the symbolics of blood. I think there are reasons to put intelligence—that most frequent target of psychologists' heredity-environment debates—on that list. Psychologists' writing on intelligence seems to bear traces of sovereign power when it allows powerful people to be made visible. Psychologists who are most invested in the category of giftedness and genius have a long history of ontologizing themselves and each other as paradigmatic of the high intelligence that they describe. Galton's example of writing about his own family when writing about the inheritance of genius is perhaps paradigmatic of this tradition.[66] Psychologist James McKeen Cattell similarly catalogued American men of science (Cattell 1906). Terman wrote an article (1917) ontologizing the high childhood

IQ of Francis Galton to support his argument that IQ tests could predict which children were bound for greatness. His student Catharine Morris Cox continued this project by using biographical materials to estimate the IQ scores of 301 historical figures (Cox 1926). Galton and his cousin Darwin show up as both the inspiration and the confirming exemplar in more recent evolutionary psychology theories about genius.[67] Intelligence testing is a domain where visibility within the disciplines that begin with the suffix 'psycho-' does not simply look like a trap that leads to normalization. Rather, the relationship between power and visibility in the IQ field appears to me to have some characteristics more typical of the symbolics of blood than of disciplinary power.

Of course, the co-construction of intelligence and sexuality through the intersecting dynamics of Galtonian and Queteletian normativity ought to explain the "stubborn particulars" of more than the Terman-Kinsey debate itself. By way of clarification, consider that most cited idea in Foucault's *History of Sexuality* (1978, 43); the hypothesis that "the homosexual" became a distinct species of person only in the late nineteenth century. This invention of the homosexual exemplified Foucault's argument (2003) as to how bourgeois norms were increasingly explained by psychiatrists with reference to sexuality, and sexuality became an anxious sign of the bourgeoisie's moral mettle, legitimating their management of complex new precarious systems of industrial capitalism. Although *History of Sexuality* did not explore them, the androcentric invention of the homosexual intersected profusely with an equally androcentric construction of the category of the genius, and these discourses echo in later stereotypes about badly adjusted geeks and nerds.[68] By describing genius as pathological, psychiatrists implicitly constructed the "normal state" of the thinking brain as not overexcited, such that gifted novelists whose works critiqued bourgeois norms became something to fear rather than admire.[69] In the late nineteenth century, genius became legible as one of degeneracy's signs;[70] Nordau's *Degeneration* opens with the observation that degenerates were not always criminals, prostitutes, anarchists, or lunatics—"they are often authors and artists" ([1892] 2006, vii)—and it continues that "Zola's novels do not prove that things are badly managed in this world, but merely that Zola's nervous system is out of order" (499). For Nordau, Oscar Wilde and his homosexuality signaled all that was wrong with the category of genius. The degeneracy theory of genius reached its zenith in Lombroso's *The Man of Genius* (1889), in which the genius is described at various turns as male, short, prone to rickets, pallor, emaciation, a cretin-like physiognomy, a large and unusual skull capacity, stammering, left-handedness, sterility, epilepsy, melancholia, alcoholism, hallucinations, moral insanity, and folie de doute.

Simultaneously, homosexuality and genius were linked to normalize the former by its association with the latter. Several early sexologists used history to associate genius and homosexuality, mirroring the explanatory scheme for eminence in works such as Galton's *English Men of Science*. Havelock Ellis considered the European history of the sexual invert to be traceable through "two classes—men of exceptional ability and criminals; and also, it may be added, among those neurotic and degenerate individuals who may be said to lie between these two classes, and on or over the borders of both" (1936, 24). Ellis claimed no originality for the idea that exceptional intellect and sexual inversion went together; Dante had recognized the link long before recent sexologists had done so (26). Magnus Hirschfeld, Ellis wrote, had constructed "two lists, ancient and modern, of alleged inverts among the distinguished persons of history, briefly stating the nature of the evidence in each case. They amount to nearly 300" (26–27).[71]

The psychoanalytic Oedipal story was also partially shaped through this discourse. Freud linked homosexuality and genius repeatedly through the concept of "sublimation," that process by which the libido was diverted into cultural achievements.[72] In *Civilized Sexual Morality and Modern Nervous Illness*, Freud described explicitly how "the constitution of people suffering from inversion—the homosexuals" often included "a special aptitude for cultural sublimation."[73] Freud's foundational psychobiography claimed that Leonardo da Vinci sublimated the sexual needs of adolescence to "take up a homosexual attitude" that "manifested itself in ideal love for boys."[74] Elsewhere, Freud described male homosexuality as a desire sublimated in the service of a higher aim, explaining why manifest, but *abstinent*, homosexuals are particularly interested in humanity.

All of these explanations of the apparent link between genius and homosexuality focused on men. Some authors noted the entomological origins of "genius" in the Latin *gignere* and the Roman notion of "genius" as a form of spirit particular to the male.[75] In an oft quoted phrase Lombroso argued that there were no great women of genius: "All the great women of genius. were men."[76] Havelock Ellis linked genius and male embodiment to explain women's limitations in the domain of musical composition; "there is certainly no art in which they [women] have shown themselves more helpless" because "the sexual sphere in women is more massive and extended than in men, but it is less energetic in its manifestations. In men the sexual instinct is a restless source of energy which overflows into all sorts of channels" (Ellis

1934, 360). Freud considered women "the actual vehicle of the sexual interests of mankind," endowed in only "a small measure with the gift of sublimating their instincts."[77] Such androcentric thinking led many American "experts" to the uncomfortable conclusion that lesbians, as inverted women, were likely to be geniuses.[78] Terman described gifted girls often as future mothers of gifted children rather than promising young people in their own right.[79] Such partiality in the sexual politics of smart men could be used to critique the pretensions of such discourses to apply to everyone.[80]

If the form of power that is involved in marking people as "genius" is not simply *disciplinary*, then what kind of power might it be? In the 1920s, Lewis Terman's used the term "gifted" to describe the intellectually precocious children whom he studied. At about the same time, the French anthropologist Marcel Mauss described the giving of gifts as an elementary form of sociality that binds giver and receiver in bonds of obligation and reciprocity because the gift is understood to remain active after it has been given. There are few gifts in Mauss's scheme that do not come with strings attached, and I want to venture that when the attribution of psychological giftedness to people is a kind of pledging that creates "binding obligations" between the authority of the psychologist and the life of the person ontologized as "gifted." When Terman described his high IQ child subjects as the embodiment of "the promise of youth," he obliged them to make good on that promise, and bound up the legitimacy of his own science with their futures. As Mauss concluded, the ambivalence of such arrangements "explains the meaning of the word *Gift* in all these [Germanic] languages—on the one hand, a gift, on the other, poison."[81] Galtonian normalization involves both entrapment and veneration.

In contrast to disciplinary power, which works to extend power over those already disempowered, "giftedness" has, in the United States, largely been gifted by academic psychologists to children from already privileged groups. Terman's study of gifted children failed to consider the possibility of Black and Latino/a gifted children, and White children have long been overrepresented in giftedness programs.[82] In the history of psychology, many minority groups, such as gay, lesbian, and bisexual people, Jewish people, African-Americans, and women scientists have used the past intellectual contributions of group members to protest the failure to recognize gifted intellectuals from their own group by Galtonian normalization.[83] My own thinking is enormously indebted to such movements, but such histories tend to rely on norms of intellectual merit that are loaded against minority groups' achievement in the first place. For example, attempts to reclaim the

intellectual contributions of overlooked women psychologists have often relied on androcentric registers of eminent psychologists for their ontologies.[84]
As Sander Gilman argues (1996), ontologizing a social group—such as Ashkenazi Jews—as "naturally" smart, can serve both to erase histories of op-
pression and to provide a basis for marginalization in the present. In the case
of "queer genius," Magnus Hirschfeld's strategy of listing "great homos in
history" tends to overlook insights—such as Foucault's—that homosexuality
and heterosexuality are modern identity categories.[85] The matters between
Kinsey and Terman give me cause to query such strategies of countering historical erasure; there is poison in the gift of historical eminence, and some
people are made to taste it more than others.

Galtonian normalization has been most successful when it is attempted
on behalf of those *unmarked* groups, who can be confused with more universal categories such as the national population. The American poet Audre
Lorde described the imagination of the United States population, and its
political interests, as characterized by a "mythical norm" of the "white male
adult Christian financially secure heterosexual." For Lorde, this figure exists
"on the edge of consciousness," distorting thinking and feeling about difference (1984, 104). Lorde's insight is important to understanding the intersecting dynamics of Queteletian and Galtonian normalization on the edges
of consciousness of what it means to be "American." Two particulars of US
ideology are important in understanding what follows.

First, consider that "IQ testing" may not be on Foucault's list of residues
of sovereign power because of the very different traditions of IQ testing in
the American and French republics. IQ is but one of many ideas that square
social inequality and egalitarian ideals by *naturalizing* differences in intellectual capacity. As John Carson details, IQ testing began with the work of
Alfred Binet in France in the late nineteenth century, but testing did not go
much further than the Queteletian project of detecting "mentally defective"
children, who might learn standard curricula slower than their age mates.
An existing system of examinations, whose humble origins Foucault traced in
Discipline and Punish, already did the work of selecting a few for rare and particular educational privilege. Carson argues that, within the United States,
intelligence testing addressed the lack of standardization within a growing
school population that had rapidly expanded westward, as well as concerns
about a much wider array of ideological dilemmas surrounding difference
and inequality.[86] Terman emerged as the figure within this movement most
concerned with the upper "gifted" end of the intelligence spectrum.[87]

Second, the distinctly American future-oriented national myth of the expanding frontier is particularly important for understanding what is most distinctive about the dynamics of Galtonian normalization, in Kinsey's work, for example. Kinsey applied the Darwinian idea that natural variation was the raw material of future progress to variation in sexual behavior. Perhaps the similarity between Terman and Kinsey's investment in the promise of talented youth is best exemplified by a speech that Kinsey delivered to the students of the Phi Beta Kappa honor society at Indiana University on June 5, 1939. Kinsey cautioned the students that many "popular judgments of normality," particularly those voiced by psychologists, often represented only "measures of departure from the standards of the individual who is passing judgment"—an admission that "'only thee and me are normal, and thee, I fear, is a bit queer'" (1971, 9). He urged the students to recognize differences as "the materials out of which nature achieves progress, evolution in the organic world," and "in the difference between men lie the hopes of a changing society." Kinsey linked his belief in their differences from each to his hope for their future: "I trust that our university has not put any standard imprint on you who have gone through it. In fact, from what I know of some of you who are the newly elected members of Phi Beta Kappa, you are a strange assortment of queer individuals, and that is why I respect you, and believe in your future" (9).

This speech shows Kinsey thinking about variation in clear Darwinian terms, as the raw material that gives hope for a better future. While Kinsey imagines this promise of youth as likely to become fulfilled by a queer unlimited non-identity, Terman's gifted study only allowed this promise to be fulfilled when talented young people were decidedly *not* queer. By drawing conscious attention to those forgotten fragments of debate, I hope to explain how that norm figure of Lorde's imagination survived the transformation from nineteenth-century "common sense" philosophy to twentieth-century American psychologies organized around intelligence and adjustment, and into the postwar era in which statistics became the register through which human consciousness was ontologized.[88]

Because of their very different investments in the normalization of sexuality, biographical writing about these two "gentlemen" has taken different forms. Galtonian normalization has had more limited and more recent effects on modern accounts of sexual alterity than intellectual alterity. People are more often taken as valid witnesses to their own pasts—and to the pasts of their allies—if they seem intelligent enough to contribute to the science of intelligence than if they seem so oddly interested in sex that they are motivated to develop a *scientia* about it. Indeed, Terman was not afraid of being historicized. He had grand ambitions that the discipline of history could be subsumed by the field of psychological testing altogether. In an article published close to his retirement, he argued that psychological tests could ultimately replace narrative ideographic understandings of individuals and groups.[89] When aged only fifty-one, he was, quite precociously, asked to contribute to the *History of Psychology in Autobiography* series.[90] He understood himself to be exceptionally intelligent, and one of his final fears was of living without "a decent IQ."[91] Terman's status as an author of a biographical paper on Galton's early IQ has had an impression on the ways that others have written about him. May Seagoe, Terman's first biographer, expressed a worry of her own powers to write about Terman, who she considered as "an expert in biography" (1975, vii).

Sexuality researchers are a different kind of strange folk, who more commonly hide from history lest their work become subject to normalization and fail to retain the capacity to shape that future. While Kinsey contributed his own anonymous sexual history to his database, he was angered when his assistant Wardell Pomeroy detected it.[92] Kinsey hoped that no biography of him would be written, insisting near the end of his life that "the progress of science depends upon knowledge. It has nothing to do with personalities."[93] Aware of the irony, Bill Condon's 2003 biopic recreated the moment when Kinsey was publicly asked—and publicly resisted—the idea that someone should make a Hollywood film about him.

And yet Kinsey's personality has incited much greater interest than has Terman's. Two biographies were written by Kinsey's colleagues Cornelia Christenson and Wardell Pomeroy in the era of the "sexual revolution."[94] In the 1990s, two further book-length biographies used archival evidence of Kinsey's extensive bisexual history to open up a wider space for the writing of Kinsey's personality. James Jones's biography *Alfred C. Kinsey: A Public-Private Life* (1997) narrated a life of deep-rooted problematic *homo*sexuality, discernable through the kind of psychoanalytic logic that Kinsey's sexology rejected.[95] The following year, Jonathan Gathorne-Hardy's *Sex: The Measure of All Things* (1998) positioned Kinsey as a "queer individual" whose bisexual sexual history was pivotal to the intelligence of his science. As much as anything that happens *within* the human sciences, these contingent normalizations of Kinsey show the joint operations of the Queteletian normalization of atypical sexuality and the Galtonian normalization of atypical intelligence at work at the very end of the twentieth century as in the very middle.

In the first decade of the twenty-first century, I felt the effects of these different forms of normalization when I visited Terman's archives at Stanford and Kinsey's archives at Indiana University. At Stanford, I was left alone to look at Terman's papers, and could pull out letters for photocopying and future analysis at will. At IU, no photocopying was permitted, and so for several days I speed-typed Kinsey's correspondence onto my laptop to carry it away. The identification of human scientists with particular aspects of human difference—such as intelligence and sexuality—continues to affect the way that twenty-first-century historians inscribe lives. In writing this genealogy of the matters between Terman and Kinsey, I have been motivated to work against these norms by dissociating Kinsey from "sexuality" and Terman from "intelligence." What comes into view from there? I offer the narratives of this book as a gift that poisons overinvestment in the measurement of human characteristics, trust in elite men to bring about a better future on our behalf, and in assemblies of support for future-oriented forms of psychology that neglect complex narratives about the past. At the start of the twenty-first century, intelligence and sexuality remain in a very uneasy relationship. Can Terman and Kinsey help us to explain the consequences of one particular set of "faulty calculations" in 1948 that gave birth to "those things that continue to exist and have value for us?"

2

Why the Gifted Boy Didn't Masturbate

Alfred Kinsey had the astounding idea that adolescent boys who experienced their first, typically masturbatory, orgasm earlier than their peers would continue to have more sex through their lives (*SBHM*, 499). Kinsey knew that in previous decades sources as diverse as Boy Scout Manuals, the United States Public Health Service, and the psychologist G. Stanley Hall had linked the wastage of men's precious fluids to weakening, a belief that Kinsey thought to be as superstitious as the ancient idea that semen came from the brain (499). Kinsey hoped that his research would refute a theory of male sexual difference in which "long-lived performance depend[s] upon the conservation of one's sexual powers in earlier years" (297). He also critiqued the most recent literature which allowed that masturbation could be normal, but only if confined to adolescence and if not carried to an undefined level of "excess." For Kinsey, masturbation was healthy, natural, and good, and questions about normality and excess were moot. Masturbation relieved nervous tension, and the adolescent boys who practiced it most frequently became "the more alert, energetic, vivacious, spontaneous, physically active, socially extravert, and/or aggressive individuals in the population" (325).

Kinsey's thesis was interlaced with his thoughts about boys' variable educational and class trajectories; college-bound boys masturbated more frequently as young adolescents but became allo-erotically active later than their peers. Collegiate men also continued to masturbate longer than their less educated age mates. Even after marriage, better educated men continued to masturbate while less educated men would "find it difficult to understand how a grown man could think of masturbating, particularly if he is married and living with his wife" (508). Working-class men moved naturally and quickly from adolescent masturbation to adult heterosexuality, while educated men had erotic responses that were subject to conditioning (157–58), made all the more likely by their reading of erotic literature while masturbating (510).

This association between early emergence of sexuality and physical health was distinctly male. In *SBHF* masturbation was surprisingly common, and emerged as *the* most reliable method of bringing about orgasm among the women who talked to Kinsey. However, Kinsey naturalized no relationship between adolescent masturbation and later sexual activity among women.[1] Rather, Kinsey's theory linking masturbation to educational success was specific to boys and men. To bolster it, he drew upon diverse findings that seemed to indicate the health of *precocity* more generally, including Terman's finding "in his study of geniuses" that "the individuals with the highest IQs were more often those who became adolescent first" (*SBHM*, 309).

Terman's response to Kinsey's ideas about sexual precocity was the first of his four distinctive criticisms of *SBHM*.[2] Terman first urged psychologists to remember the distinction between their own, more cautious, attempts to measure personality traits and Kinsey's casual impressions: "psychologists who have found it so difficult to devise reliable measures of such personality traits will be interested in these ratings" (1948, 452). Terman also spotted that Kinsey's own data argued against the thesis; sexually precocious active men were less active in old age (452). As Terman developed this critique, his argument moved onto shakier ground as he made claims as to whether men could or could not accurately recall their earliest sexual experiences. Kinsey had used men's accounts of their first orgasm to date the onset of their adolescence, and had used memories of physical changes associated with adolescence to suggest when an orgasm *could* have occurred among those men whose first orgasm occurred later. Terman understandably judged that "the rules here set forth are involved and 'iffy.'"

But Terman's critique of Kinsey relied upon an iffy assumption of its own:

> The reviewer doubts whether any man ten or twenty years beyond adolescence could give more than a wild guess as to his own age at first ejaculation, at first appearance of pubic hair, or at onset of rapid growth. It is not merely a matter of memory, each of these signs of adolescence makes its appearance gradually. (1948, 451)

Terman even suggested that "such judgments would seem to call for a kind of occult insight that most people don't have" (452). In other words, rational men not only did not, but *should* not, remember these sorts of things. A science of human sexuality built from such reported memories would have something of the night about it. It could only be iffy, at best, and might, at worst, be occultish.

What does this exchange begin to reveal? In this chapter I want to consider this exchange as a minute turning point from which a broader landscape can be glimpsed; the mutually constitutive history of masturbation as a problematic embodied relationship to the self and the emergence of a disembodied psychological science of measurable differences between other people. Beyond psychoanalytic circles, few psychologists wrote much about masturbation prior to 1970. Since then, psychologists have worried both about people deemed to masturbate too little (such as married women) and those deemed to masturbate excessively (such as wards of the state in prisons, hospitals, and other institutions).[3] Terman is not atypical. Although my reading of his works has not been exhaustive, as far as I can tell he wrote little about anyone else's masturbation and nothing about his own. In the book that Kinsey cited—Terman's first volume on his study of gifted children—Terman noted only that "seven histories of masturbation were reported to the physicians, including 4 boys and 3 girls. The data on this point are probably very incomplete" (1925, 210).

In contrast, Kinsey's volumes have been rightly considered by historians as pivotal to the twentieth-century repudiation of Enlightenment concerns that masturbation is a diseased, weakening, unhealthy habit.[4] While Kinsey's analogy between sexual and intellectual precocity and Terman's response to it have not been discussed in this context, such a discussion might help to account for that unreflexive character of American psychological science, oriented to studying "others," which feminist historians have described as a distinctly male form of subjectivity.[5] It is telling that Terman appealed to measurable personality traits to dismiss Kinsey's claims about the characters of his participants. Historian of psychology Kurt Danziger has tracked a shift in the psychological research literature of this period away from studies of named introspecting observers towards standardized tests conducted on groups of anonymous subjects whose scores are aggregated.[6] This shift—from "self-regard" to "other-regard" in psychology—happened first in applied psychology journals and was led by IQ researchers,[7] and the Stanford psychology department under Terman's leadership from 1921 onward exemplified the shift.[8]

The exchange also points to a third figure whose influence on this debate will be the starting point of my discussion. Kinsey identified psychologist

G. Stanley Hall (1844–1924) as an important source of anti-masturbation influence in early-twentieth-century American culture, and Hall was also Terman's first postgraduate mentor. While historians of masturbation have rarely attended to North American academic psychology, they have often alighted on Hall as an important late spokesperson for anti-masturbatory Enlightenment thought.[9] Accordingly, I will turn to Hall next to consider both his specific influence on Terman, Terman's reaction to that influence, and Hall's broader influence on American culture by way of explanation of Terman's silence about the masturbation of the gifted children and his response to Kinsey's breaking of that silence.

Psychologists Forgetting Masturbation

Anti-masturbatory Enlightenment thought has been historicized from the popular tracts of the early eighteenth century, through the medical writings of Tissot in the mid-eighteenth century and his nineteenth-century followers, to early-twentieth-century writers, including Hall's magnum opus *Adolescence* (1904).[10] The mid-nineteenth century brought new reasons to fear that male sexuality might stray from the developmental path to its heterosexual telos. "Bachelors" were a new cause of anxiety as more men spent time earning wages prior to marriage.[11] Working-class men who left home to work in urban centers received equally passionate advice against masturbation if they boarded at the YMCA. The disease of "spermatorrhea," evidenced by leakage from the penis, occupied the minds of some middle-class men, often doctors, who feared that their seminal emissions evidenced past and future weakness.[12] Patents for chastity devices were issued in the United States with greater frequency in this period.[13] The context of Hall's own "adolescence," avant la lettre, was one that was structured by a profound fear of male masturbation quite foreign to our own post-Kinsey era.

Because of such new efforts, Foucault (1978) described the Victorian period as one organized not primarily by the "repression" of sexuality but by the "production" of new discourses about sexuality's nature, causes, and consequences. Hall fits Foucault's account; he wrote passionately about masturbation in others and was also that irrational adult male figure of Terman's imagination—a man who freely confessed his adolescent sexual experiences (1923, 131–32). In autobiographical writings Hall described harsh admo-

nitions from his father, who led him to think of his own genitals as naturally dirty and of masturbation as likely to make his nose drop off (131–32).
Hall struggled to avoid the habit and occasionally slipped up. Typical of
nineteenth-century college men, his sex education included not only his father's warnings but also the "common sense" philosophy routinely taught at
American universities in which elite male life was a heroic battle of the will
over animal nature.[14]

Hall is a figure who is difficult to locate in the history of North American
psychology without using the term "first."[15] Delicate readers may wish to
look away while I indulge the impulse in this paragraph. Hall was, in 1878,
the first to receive a Ph.D. in psychology on American soil (from William
James). Hall was also, in 1887, the founder of the first journal of psychological research in the English language, the American Journal of Psychology. In
1892, he became the first president of the American Psychological Association, and by 1889 he held the chair of the first psychology department in the
country, at Clark University. Founder also of the journals Pedagogical Seminary, in 1891, Journal of Religious Psychology, in 1904, and Journal of Applied
Psychology, in 1915, Hall had by 1898, supervised thirty of the first fifty-four
American Ph.Ds in psychology.[16] He hosted Freud's 1909 visit to the United
States.[17]

In short, Hall was not an occult shadowy figure in the history of American psychology. And yet, much like the account of adolescence as a developmental period of "storm and stress" promoted in his own work, his contributions are often historicized as an awkward moment of growing pains for
the discipline, best remembered for the mature forms that developed out of
them than for the ungainly shape of Hall's own thoughts. By analogizing the
development of the human individual and the emergence of an intellectual
field, I am drawing analogies in a way that is characteristic of Hall's own
work. Aptly located by his biographer in "the late Victorian, early Darwinian era,"[18] Hall espoused a recapitulatory theory of evolution, in which the
history of the growing person became a metaphor for the purported history
of the species and the intellectual development of a culture.[19] Indeed, by
the mid-1890s, Hall was using masturbation as a metaphor to specifically
critique the futility of idealist philosophy.[20] The "genetic psychology" that he
developed in this period understood the materiality of life and human consciousness to be linked, such that ongoing processes of evolution could be
studied through empirical research on living children, who re-lived that evolutionary drama in their individual lives. Free from later tensions between

science and Christian belief, Hall was explicit that the hand of a Christian God was the primogenitor of the material world, and that American Christianity was the culmination of an evolutionary process. In Hall's writings, spirit and flesh were neither identical nor estranged, but they frequently met in metaphor. As a result, contemporary psychologists invested in the modern literality of their own scientific discourse might feel a twinge of embarrassment in sharing a disciplinary space with Hall's verbose, prolific, metaphorically rich, scholarly, statistically light, and historically oriented writings—writings that are awash with concerns about religious conversion, sexuality, and outmoded models of evolutionary change. In other words, to read Hall's words now in the twenty-first century is to notice their marked difference from the technical genre of experimental writing in psychology which has often been deemed masculinist by virtue of its lack of reflexivity.

In the early 1880s, Hall was the central figure in the development of an American science of pedagogy. Responding to new legal mandates for compulsory primary education, Hall's "child study" movement brought into being new networks of researchers that involved academic psychologists, teachers seeking professionalization, and middle-class women with scientific education and ambition but who were unlikely to ever receive university appointments.[21] Hall was an educational romantic who held that children ought to be educated through direct interaction with nature away from the fast pace of urban life. By the mid-1890s, Hall's psychological fellows were increasingly alienated by his intolerance of their philosophical writings and his arrogant claims to be training the only psychologists of merit in the country. When Hall wrote *Adolescence*, his science was already in danger of being superseded by antagonistic developments in American psychology. While Hall advocated empirical psychology, he failed to keep pace with the quantitative impulse that continued to separate psychology from its philosophical moorings. By the start of the twentieth century, state normal schools were drawing their faculty from the statistically oriented Teachers College at Columbia rather than Hall's department at Clark.[22] *Adolescence* was an attempt to reclaim Hall's position at the center of American psychology, a discipline that was beginning to leave his vision of the future of civilization behind.

As this is a chapter about the possibility of psychologists putting their guilty secret pleasures on the record, let me gamble one of my own. Hall's invective against adolescent male masturbation in *Adolescence* is among the most passionate writings, by a psychologist, that I have ever read. As sociologist Lisa Moore (2002) has argued, complex competing definitions of hegemonic masculinity meet in scientists' writings about semen, and Hall's writings on adolescent masturbation are no exception. Hall's chapter describes masturbation, and the possibility of its eradication, as "one of the richest scientific quarries opened up by the new psychology" (1904, 432). Primarily concerned with male masturbation, *Adolescence* allows that adolescent girls can masturbate, but it is typical of Victorian male "expert" views of women as barely moved by sexual desire at all and oriented instead towards the pleasures of motherhood.[23] For Hall, the power of the will that struggled against masturbation had a vexed relationship to the materiality of semen. All biological matter housed something like "Schopenhauer's basal will to live" whose "all-dominant impulse is to progressive self-expression," and which originates in God himself (1904, 411). This will was made flesh in "spermin" (chemical formula $C_5H_{14}N_2$), which existed within semen, purified the blood, freshened the nerves, and increased self-feeling. Spermin was "perhaps the highest and most complex of things in the physical world" (442), but it became wasted by masturbation and excessive sex. Thus the urge to masturbate and the force of character required to resist that urge shared a distinctly male materiality. The drama was of supernatural proportions, for when resisting the temptation to self-abuse, "the young man is fighting the hottest battle of his life with the devil solitary and alone" (458).

For Hall, White men especially felt the temptation to masturbate, and especially resisted it. Writing after the *Plessy v. Ferguson* decision which institutionalized racial segregation for over half a century, Hall described how his Black compatriots should not be taxed beyond the level of technical education.[24] Against a majority of educational thinkers of his time, he also endorsed gender-segregated education and held that women's reproductive capacities might be harmed by strenuous schooling.[25] While White men were the pinnacle of evolution, evolution was also responsible for a threat to civilization; masturbation was a direct result of the evolution of the human hand (1904, 438) and—as Kinsey would later echo—the postponement of marriage among civilized people (452). Because masturbation squandered the potential for further human advancement, it was "one of the very saddest of all the aspects of human weakness," and "the most perfect type of individual vice and sin" (452). Tragically, in regard to masturbation "civilized man is on the whole, to say the least, no better, if not far worse, in this respect than his savage brother" (435).

Because civilized men were characterized by both the urge to masturbate and the capacity to resist that urge, Hall found reason to fear for the habits of both intellectually precocious and intellectually retarded boys. Hall might have agreed with Kinsey that precocious scholars were likely to also be precocious masturbators. But Hall would have explained the correlation as an effect of a naturally compensatory system in which the premature emergence of the intellect leads to weakness, and not as the natural co-emergence of two related signs of a vigorous constitution. Hall figured the promise of the gifted intellect as precarious and masturbation as likely to compromise the futures of both bright and stupid boys. Boys of "precocious mental development" might develop the habit "in complete ignorance of its meaning." "Wherever children mature early in mind there is special danger of a wrong direction, and therefore need of all the methods of control" (436). This White man's burden was likely to become unstable also because mental overwork created "that irritable weakness of the nervous system which makes the rankest ground for onanism" (436). Hence, Hall wrote of "perversions of the power of imagery in youth" that generated both shame about having masturbated and sexual fetishism through the introduction of non-natural fantasies (411). On the other hand, stupidity was no safe haven against masturbation either, "because it is so dangerous, and liable to occur in individuals who lack stamina, it has its octopus-grasp in nearly all institutes for the defective classes" and is "practically ineradicable in institutions for the feeble-minded" (434).[26]

Hall was both critical and reiterative of Victorian theories about the harmful effects of masturbation, and his attitude to masturbators was closer to pity than contempt.[27] He was concerned for those "victims" who were "sometimes sold placebos or harmless drugs, and if made well are really healed by faith or by alleviation of worry," particularly because "most of the apparatus also is sold at exorbitant prices that consume the savings of years and sometimes prompt to theft" (461). In a post-Kinsey era, Hall's recommendation to apply the power of the will to overcome the urge to masturbate seems as quackish as the cures that he warned about. However, Hall wrote *as* a masturbating subject, that occult figure who recalled his own solitary struggles with the devil, and who confessed them in writing (1904, 461–62). As such, his passionate writings might be more usefully remembered as continuous with our own unresolved emotional relationships to the stigma of masturbation, and with the limits of pity as an emotional response to sexual stigma, in Hall's time and our own. Writing off Hall's work as an awkward growing pain of the now mature scientific discipline of psychology would tend to render such insights rather "occult."

Disembodying Smart Boys

While Hall espoused several exclusionary ideologies, they were not well translated into his educational practices at Clark University. In spite of his opposition to coeducation, and his belief in the inferiority of the female intellect, many smart women passed through his ambit.[28] In spite of his explicit racism, he graduated Cecil Sumner in 1920, the first African-American to receive a Ph.D. in psychology in the United States.[29] Hall's theory that a hardy constitution was required for the intellectual demands of higher education has suggested complicity with the increasing prejudice against tuberculosis patients of the 1890s.[30] Yet, Hall often admitted tuberculosis sufferers for advanced study at Clark.[31] In the year before *Adolescence* was published, he admitted the young Lewis Terman, for example.[32] Lewis Terman grew up in a large Indiana farming family. He trained as a teacher, and suffered the first of several bouts of tuberculosis at age nineteen that would haunt his long life (Seagoe 1975, 8). In 1899, he married Anna Minton, who later bore two children, Helen and Fred. Lewis Terman returned to education to study psychology at Indiana University, one of the first departments in the country. There he wrote essays on "degeneracy" and "the great man theory" before moving to Clark to begin his Ph.D. studies.[33]

Hall's Monday night seminar left a lasting impression on Terman. The seminar was a model of collective intellectual engagement that he would replicate later at Stanford.[34] Years later, Terman remembered the physical stress that intellectual engagement with Hall would take on his own fragile nerves:

> Dr Hall usually started the discussion off with a few deceivingly generous comments on the importance of the material that had been presented, then hesitantly expressed just a shade of doubt about some of the conclusions drawn, and finally called for "reactions." . . . When the discussion had raged for thirty minutes to an hour, and was beginning to slacken, Hall would sum things up with an erudition and fertility of imagination that always amazed us and made us feel that his offhand insight into the problem went immeasurably beyond that of the student who had devoted months of slavish drudgery to it. . . . At the end of each session, I always went home dazed and intoxicated, took a hot bath to quiet my nerves, then lay awake for hours rehearsing the drama and formulating the clever things I should have said and did not. (Terman 1932, 315–16)

In Hall's theory and practice, a healthily stimulated intellect required a robust constitution. In contrast, Terman's science of measuring Intelligence would *disembody* intelligence, making such relationships between body and mind less pertinent.

At the start of the twentieth century, few psychologists concerned themselves with measurable intelligence. By the time that Terman arrived at Clark, James McKeen Cattell's tests of sensory acuity and response speed had failed as means of measuring individual differences in academic achievement.[35] In 1904, the British psychologist Charles Spearman argued that scores on mental tests were sufficiently correlated that they suggested a single underlying "General Intelligence."[36] A year later, the French psychologist Alfred Binet published a new form of graded test that aimed to distinguish normal French children from those who ought to be placed in special education.[37] The materialization of a property called "Intelligence" was in the air once again, and several of Hall's students from the 1900s were to make material contributions to its development as *the* theory of intellectual differences in American psychology in the coming decades.

Two articles that Terman published while a student at Clark demonstrate Hall's influence and Terman's move away from it. "A Study of Precocity and Prematuration" (1905) was published in Hall's own journal, *Pedagogical Seminary*. The article began by critiquing the "loose talk" about the concept of "precocity," which Terman defined as "development in advance of some norm" (1905, 145), Terman's distinguished natural "precocity" from unnatural "prematuration." The latter resulted from "forced culture" as exemplified by a plant pruned to encourage its growth, an adolescent encouraged to act like an adult, or an attempt "to initiate into the harmonies of Mozart, minds that find more edification in the rattle of tom-toms" (148). Terman did not oppose "prematuration" and "precocity" entirely consistently within this article, but he repeatedly argued against the normalization of naturally early-maturing children, while maintaining that the instinctive dread of human difference remained a reasonable response to "forced acceleration."

The article moved away from the kind of Queteletian scheme of normativity evident in *Adolescence* in which both early and late development were seen as potentially problematic. Terman put forward a more Galtonian vision of difference in which unusual early development could be *naturally* precocious, unalarming, and even valuable. To be sure, Terman recapitulated Hallian views on the nervousness of gifted children, the dangers of education that fails to strengthen the will and character, and the abnormality of sexually precocious children. However, he also questioned the urge to normalize children who seemed to develop particularly early:

> It is so convincing to talk of the "normal" and the "abnormal." It appeals to our instinctive dread of being unlike our fellows. To designate a quality or character as "abnormal" is, in the popular mind, to condemn it. Hence, children of unusually rapid development have sometimes been called "monsters."
> (1905, 147)

This normalization of natural precocity found a limit in the domain of sexual development: "no other phase of precocity is so important as that related to the premature development of the sexual functions, and no other is so difficult to treat" (173). Consistent with Freud's contemporary arguments, Terman argued that norms of childhood ignorance led people to underestimate how common childhood sexual precocity might be.[38] Like Hall, Terman voiced concern that perversion could result from precocious sexuality unchecked by mature reason. "The evidence is convincing that a good portion of those of perverted sexuality acquire their practices long before they have reached an age when most people regard oversight as necessary" (177). The flourishing of inversion in gender-segregated schools, a curriculum that bored students with rote learning, and the opportunities for the imagination to stray to perverse corners of the mind were all described as risks of an educational system which forced an unnatural culture on growing children. The Hallian solution was clear: "The problem is largely to strengthen the volitional powers before the onset of puberty. Only a strong will can guide the human bark through the storm and stress of adolescence" (182). For Terman, this will was located in the motor areas of the brain, and the neglect of physical education for classroom instruction made the schoolboy sluggish, nervous, and prone to sexual precocity and "an over-functioning of the associative centres" (182).

In his subsequent dissertation, "Genius and Stupidity: A Study of Some of the Intellectual Processes of Seven 'Bright' and Seven 'Stupid' Boys" (1906), Terman moved away from Hall's influence, and the ontology of the boys' brightness in that work rendered the possibility of an association with sexual precocity much less likely.[39] Terman's dissertation was prescient of the methods of statistically comparing artificially created groups that historian of psychology Kurt Danziger (1987) described as "neo-Galtonian," in which psychologists compared the scores of subjects who were grouped according to an arbitrary measure of the psychologists' own design. Specifically, Terman used teachers' judgments to construct the two groups of "bright" and "stupid" boys, "all from respectable, middle-class homes" (1906, 317). He subjected both groups to a battery of mental and motor tests that ranged from running up stairs, to learning the rules of chess, to the interpretation of fables. The bright boys excelled on the mental tests, but the stupid ones did better on the motor tests. The study served to disembody brightness, severing that connection between smart minds and healthy bodies which had been causally connected in Hall's account of the dangers of masturbation to intellectually oriented White boys.

Terman's smart boys did not need the housing of a strong body, and the paradigmatic objects that their brightness illuminated were not natural either. Their performance on artifactual games like chess and contrived puzzles revealed their smartness. These two early papers ontologize very different kinds of bright boys. In the first, dangerous "over-functioning" associative centers suggest the risk of sexual inversion. In the latter, multiple speedy mental associations are the very stuff of a less embodied brightness that appears unlikely to signify sexual precocity at all. In this latter paper, informed by British psychologist Spearman's statistical construct of "general intelligence," Terman defined the difference between brightness and stupidity in terms of the number and speed of mental associations. While Terman acknowledged "mechanical," "artistic," and "scientific or philosophical" forms of intelligence, the scientist and the engineer, rather than the artist, were his paradigms of brightness. The speed of the "associative centers" of bright boys' minds had become a cause for optimism rather than alarm.

The Morality of Intelligence:
Stupid Prostitutes and Gifted Paragons

Terman's Galtonian scheme was not simply "unreflexive"; it afforded new implicit and explicit ways for psychologists to talk about themselves, each other, and people that they admired. The IQ testing movement, like the bright boys in Terman's dissertation study, was itself precocious, imaginative, and worked to transform the outer world and to solve the problems perceived to lie within it. The wide-ranging movements by professionals in diverse institutions to make matters of social concern into objects of scientifically informed policy, and the growth of professions fueled by such aspirations, characterized the "progressive era." During this period, American experts pledged to solve ideological problems and to address political corruption in a rapidly urbanizing immigrant-based, post-slavery democracy.[40] The IQ testing movement is a case in point. In the first two decades of the twentieth century, "intelligence" became an everyday word in the United States.[41] Henry Goddard first demonstrated the use of Binet's test to detect "feebleminded" people in 1910.[42] Several American psychologists worked with Binet's testing methods and others, in the six years afterward.

Within this context, Terman's 1916 report of the Stanford-Binet test of intelligence in his book *The Measurement of Intelligence* is remembered as setting "the paradigm for virtually all the written versions that followed,"[43] propelling Terman to fame and fortune,[44] and marking "a fundamental divide in the American history of intelligence."[45] The test has been critiqued by psychologists and historians for the racism of its norming procedures (and its consequent attribution of higher intelligence to Whites), its failures to recognize the intelligence of people who might respond to its items by ironizing them, and the inability of intelligence researchers to adequately define the construct that the tests appeared to measure.[46] Consistent with the privileging of the *scientific* intellect, Terman drew on an analogy from the physical sciences to justify how intelligence testing could proceed intelligently without a clear definition of intelligence:

> As Stern points out, electrical currents were measured long before their nature was well understood. Similar illustrations could be drawn from the processes involved in chemistry, physiology, and other sciences. (1916, 44)

In other words, the version of "intelligence" that Terman ontologized as "IQ" was highly particular, as all accounts of this concept have since been.

Histories of the IQ movement have emphasized how testing emerged from institutions for the feeble-minded, psychiatric institutions, and schools, and that engineering and medicine provided conceptual language for the growing mental testing movement.[47] However, social and mental hygienists who concerned themselves with prostitution, venereal disease, and sex education were also interested in the measurement of the growing intellect, and their work also influenced the definition of Intelligence that Terman presented in *Measurement*.[48] Therein, Terman proposed that Intelligent people were those who were best able to maintain a hygienic life in a volatile democracy. The Stanford-Binet test did not measure emotions or the will "beyond the extent to which these naturally display themselves in the tests of intelligence" (1916, 48). Nor was it a test of "those emotional and volitional aberrations which are concerned in such mental disorders as hysteria, insanity, etc." (48). However, "defects of intelligence, in a large majority of cases, also involve disturbances of the emotional and volitional functions" (49). Terman stated the relationship most clearly when answering a question that he set for himself: "why do the feeble-minded tend so strongly to become delinquent?"

> Morality depends upon two things: (a) the ability to foresee and to weigh the possible consequences for self and others of different kinds of behavior; and (b) upon the willingness and capacity to exercise self-restraint. That there are many intelligent criminals is due to the fact that (a) may exist without (b). On the other hand, (b) presupposes (a). In other words, not all criminals are feeble-minded, but all feeble-minded are at least potential criminals. That every feeble-minded woman is a potential prostitute would hardly be disputed by any one. (1916, 11)

Contemporary psychologists may be surprised that Terman linked Intelligence to morality, or that *prostitution* figured the relationship between feeblemindedness and crime. However, "expert" men often used the term "prostitute" in the early twentieth century to refer to young women with desires that were not contained within heterosexual marriage rather than specifically to sex workers. Prior to the emergence of a visible "girl problem" created by respectable young middle-class women's irrepressible desires for sex with soldiers in World War I, the term "prostitute" often served to police the degree to which such women could enjoy the freedoms opened up by urban life in the United States.[49] In the years prior to *Measurement*, several psychologists investigated the hypothesis that "prostitutes" had lower IQ scores than other women. Terman's IQ test found one of its first uses in the Sonoma State Home for the Feeble-Minded in California, where 825 incarcerated women were categorized as "idiots," "imbeciles," and "morons."[50] Goddard concluded that most "wayward girls" were feebleminded and that "it is these weak-minded, unintelligent girls who make the white slave traffic possible" (1923, 14). Terman also drew on research conducted by Jean Weidensall under the authority of Katherine Davis as superintendent of Bedford Hills Reform Prison. Weidensall (1916) concluded that prostitutes' IQ scores were bimodally distributed; some were feebleminded, but not all.

This specification of the attributes of prostitutes could be considered a form of Foucaultian "disciplinary power," individualizing poor women who were already powerless, so that they might be yet more tightly controlled, and in the interests of their own moral salvation. However, the Galtonian scheme of normativity also allows for positive descriptions of atypical people who embody ideals. Within the intelligence testing movement, Terman is *the* figure most persistently interested in the "upper end" of the intelligence spectrum. Terman's writings on gifted children—like his comments on prostitutes in *Measurement*—show a consistent concern to associate sexual morality with high intelligence. Terman was aware of the literature, reviewed in his own 1904 paper on precocity, which associated the early development of the intellect with the risks of masturbation and sexual inversion. In 1915, he directly asked the question "are children of genius usually defective or queer?" and responded in the negative with other-regarding science on thirty-one children who had scored above 125 on the Stanford-Binet IQ tests (1915, 534). When their teachers were asked to describe their moral traits, most were described as having "favorable moral traits only." Only one, who was described as "sexually abnormal and vicious," was considered by Terman to show traits that were "really serious" (535). In *Measurement*, Terman similarly described several gifted children who were neither nervous nor queer, asserting that the "feeble-minded genius" was a creature of myth, not an object of science (1916, 100–101).

In 1919, Terman began a landmark study of gifted children supported initially by Stanford University and later by the Commonwealth Fund, which became a precocious advance in the study of life-span development in the new discipline of psychology. As sociologist Leslie Margolin has detailed, Terman's work was central to the construction of the category of "gifted children" as a new "way of being a person" constructed around an ideal of hard work and classroom comportment.[51] Jessie Chase was a Stanford student who interested Terman because of her high own IQ, and she was first charged with following up thirty-one children of high IQ that Terman had identified in his earliest studies with the Stanford-Binet.[52] In 1921, she organized the six-week-long fieldwork training of four psychologists, Florence Goodenough, Helen Marshall, Florence Fuller, and Dorothy Yates, who were to find gifted children in California's cities' schools. Teachers were asked to propose their brightest and youngest students, who were given a pencil and paper IQ test. Initially Terman had planned to select the top 5 percent of this group for Stanford-Binet testing. However, extracting giftedness from classrooms was harder than Terman anticipated. To keep the project going, 10 percent of the older children and 15 percent of the younger children had to be selected for Stanford-Binet testing. Three hundred and seven girls and 354 boys were ultimately included in the gifted cohort.

Given the overrepresentation of White, middle-class, and urban children in the project, and the almost equal number of boys and girls, the early phases of Terman's gifted study have been more frequently critiqued as more blatantly racist and classist than as sexist.[53] However, the study also remained thoroughly androcentric from the organization of intelligence needed to carry it out. Like Hall's child study movement, Terman's gifted project relied on smart women in the field to turn children into data, but Terman often failed to recognize the intelligence of the women on whom he relied. Dorothy Yates critiqued the invasive questions she was demanded to ask of parents while securing their cooperation and participation in the study.[54] Helen Marshall reported that Terman's definition of genius as an IQ of 140 was arbitrary and convinced him to include one girl who fell below this threshold.[55] Florence Goodenough tried to broker her status as a researcher on the project to enroll in classes at Stanford, a move which Terman repeatedly blocked on the grounds that the number of women allowed to take such classes was capped. In other words, these women developed "situated knowledges" which anticipated later critiques about the degree to which IQ test scores depended on rapport and functioned to reinforce hierarchies of educational privilege.[56]

Genetic Studies of Genius I (GSG I) also announced a break with Terman's Hallian roots. Its opening pages directly repudiated Terman's earlier paper on the evils of prematuration. A bolder, more future-oriented account of gifted children was now offered. Rather than being the victims of modernization, undiscovered gifted children were to be rescued from lives of mediocre education by the educational engineering of IQ testing. If nurtured correctly, they would progressively move society forward through their inherited superior talents. In the modernist fashion common to psychology, Terman described the psychology of genius as having a long past and a short history. Genius had been studied by the ancient Greeks, but its scientific study awaited the undoing of such false beliefs as the idea that "the Great Man" is qualitatively different from others, that "intellectual precocity is pathological," and democratic sentiments which encouraged "an attitude unfavorable to a just appreciation of native individual differences in human endowment." The science of genius had to wait particularly for "the tardy birth of the biological sciences, particularly genetics, psychology, and education" (1925, v). Both the gifted children and the psychological study of giftedness seemed deserving of the future.

In this study, Terman did not attend to the masturbation of the gifted children, but he was not disinterested in their sexual precocity. Rather, empirical proof of the normality of their physical development was part of the normalizing project. Psychologist Bird Baldwin investigated one hundred fifteen of the boys aged nine or older and all of the girls aged ten and older for the presence and the "appearance of a kink or twist" in their pubic hair. For boys, the emergence of the hair dated their pubescence and a visible kink dated puberty, and the gifted boys appeared to be maturing earlier than a cohort of children assessed in an earlier study of New York City children.[57] Terman also found that the gifted girls began menstruating earlier than other girls. These findings showed that precocious intelligence was not bought at the expense of delayed physical development, as Hall's compensatory logic would have predicted. Yet, in spite of this invasive anthropometry of children's bodies, Terman's first book on the gifted contains only the single sentence on masturbation cited earlier in this chapter.

When we consider the shifting discourses connecting masturbation to psychological adjustment and Terman's obvious desire to avoid the sexual stigma attached to the gifted, this silence appears to have been a shrewd strategy. In the 1920s, the risks of linking growing children with masturbation were multiple. While the field of psychology had moved away from Hall's theory by the 1920s, Hall's work remained influential within the mental hygiene movement. Hall's 1909 Clark University Conference on Child Welfare captured the attention of the growing number of social hygienists in New York concerned with the spread of vice, thought to be the almost natural consequence of urbanization. Jeffrey Moran has astutely described the cadre of sex educators that developed after this conference as caught on the horns of a dilemma between "teaching young people proper information about sex before their minds were thoroughly debauched" and avoiding "the possibility that this education would itself arouse precocious interest in sexual matters." However, "between the need for timeliness and the dangers of suggestiveness lay an exceedingly narrow path" (Moran 2000, 39). In other words, the development and education of children's sexual knowledge remained severely limited by Queteletian fears about both delay and improperly precocious development. In contrast, the practices of IQ testing represented any advance of "mental age" over chronological age as an unalloyed gain, a precocious development toward a more advanced future. Occasionally sex educators looked with envy at the success of the IQ movement, hoping for a Binet test of sexual knowledge that never materialized (Moran 2000, 48).

On the other hand, by the 1920s the unfeasibility of sex educators' goals of eliminating the desires of boys in order to make them similar to the passionless girls of Victorian imagination became increasingly obvious. A counter-discourse that attributed the purported ill effects of masturbation to the neuroticism caused by trying to suppress the habit was gaining currency. This emergence was not a sudden thing; even Hall had wondered if some of the ill effects attributed to masturbation were not psychological effects of guilt occasioned by masturbation. Influential works by sexologist Havelock Ellis and psychoanalyst William Stekel began to attribute the purported negative effects of masturbation to the anxiety experienced by fearing and avoiding the habit. As Kinsey would later point out, the 1920s literature increasingly referenced *excessive* masturbation without ever really quantified the amount of masturbation that a growing boy should be engaging in to consider himself normal.[58]

In sum, available discourses could have troubled the normalization of the gifted boy who masturbated too little or who masturbated too much. The first boy risked being considered as neurotic, and the second as vicious and queer. The normalization of this strange group of children was best secured by knowing very little about their private habits. If any of the gifted boys were fighting a solitary battle with the devil, then such boys were truly alone. Not even those peculiar psychologists who concerned themselves with such boys' emerging pubic hair, and whose other-regarding science had rendered them "gifted" to begin with, preferred not to know. Kinsey's precarious analogy between variability in intellectual and sexual precocity interrupted this silence. Terman's response was to attempt to reestablish it by positioning Kinsey's claims beyond the legitimate science of measurable individual differences and by insisting on the injustice of any science that would require a grown man to report on his adolescent history of self-abuse.

Like Kinsey, I am intrigued when I look back at Terman's interest in the timing of the onset of adolescence among the gifted children. But the limits of Terman's empirical gaze, and the precariousness of the normalization of the gifted children that he accomplished, intrigue me most of all. Terman's science of individual differences was not simply an ever-expanding form of disciplinary power that sought to know in ever increasing detail. It was also characterized by deliberate strategies of silence that directed the empirical gaze in some directions and closed down others. As we will see in later chapters, this was not the only place where Terman obscured the possibility that giftedness could signal sexual alterity, or the only place where Kinsey's *SBHM* threatened to muddy the waters that Terman had stilled.

Nor has this little matter between Kinsey and Terman been put to bed by the enormous growth of psychological science between their time and our own. Kinsey and Terman might have given different advice as to how to know whether a thirteen-year-old working-class boy was bound for college or not. One would have given him an IQ test and the other would have checked the bed sheets. While Terman's answer to this question appears more evidence-based, I cannot square my own subjective recollections of growing up with male-typed embodiment with Terman's assumption that the first ejaculation is always experienced as an unsudden thing, akin to the growth of limbs or of body hair. Nor does the modern *scientia sexualis* of recent decades provide an easy resolution. Carolyn Halpern and her colleagues (2000) found that men disclosed their adolescent masturbation in interviews less often as adolescents than as adults; adult men, *in general*, are not necessarily poorly positioned to recall this event to a sex researcher. Indeed, James Jones's biography of Kinsey demonstrates why the stigma of the near-universal practice of adolescent male masturbation still matters. Assuming that a brush that was found in the Kinsey home was a masturbatory aid that belonged to the adolescent Alfred, Jones wrote that "boys bury more than their treasures; they also hide secrets—dark, fearful things about themselves they do not want others to know" (1997, 82). This "discovery" of Kinsey's purported secret begins James's larger narrative which pathologized Kinsey as a repressed homosexual and self-hating masochist, whose conflicts limited the rationality of his sexual science.[59] In the next chapter I will examine such relationships between biographical accounts of the private and the ontological status of these scientists' work returning to Terman's account of the "ideal husbands" that "bright" boys might someday grow up to be.

3

Less Than Ideal Husbands

In the previous chapter, I described how Terman's silence about the masturbatory habits of gifted children muted questions about any possible relationship between high intelligence and the emergence of precocious male sexuality. This chapter concerns the *second* of Terman's criticisms—regarding Kinsey's claim that rates of premarital intercourse were stable across the generations—and it centers on the less-than-ideal husbands that gifted boys might grow up to be. Terman's 1938 book *Psychological Factors in Marital Happiness* (hereafter *PFMH*) is less well known to contemporary historians than his IQ work. But Kinsey read *PFMH*, and, writing his own *SBHM*, he wrote to Terman with a question about his work:

> I have commended you in my forthcoming book for having made the breakdown of the population which you did in the volume of Factors in Marital Happiness. Our own study, however, has shown that there are considerable differences in patterns of sexual behavior at different social levels, and it is probable that you would have gotten different results for your population if you had had a further break-down on the basis of the number of years of schooling the subject had, or on the basis of some other classification such as sociological classification showing social levels involved.[1]

Just prior to writing this letter, Kinsey had met Olga and Quinn McNemar, a married couple of psychologists who were among Terman's Stanford colleagues. Olga had been a research assistant on Terman's marriage project. Quinn had been first Terman's graduate student and was now his faculty colleague.[2] Kinsey continued by asking whether "Mrs. McNemar, or someone else under your direction, could re-calculate some of the data in regard to sexual activity for the 71 percent of your population which went to college," and he predicted that the resulting statistics "would be quite comparable with what we have gotten in our six and eight-way breakdowns in which educational level and social class have been involved."[3] Terman did not oblige Kinsey's request for more complicated analysis but replied instead that the relevant data had already been destroyed.[4] In his original book, Terman had compared age cohorts within his largely middle-class sample, found premarital sexual experiences to be increasing, and raised the alarm that "at this rate of increase, intercourse with future spouse before marriage will become universal by 1950 or 1955" (*PFMH*, 321–22). This change was not one that he welcomed. In that same study, Terman had found that virgins and their spouses had happier marriages than their more experienced counterparts (325).

Contrary to his letter, Kinsey did not ultimately commend Terman's marriage book in his own volume at all. Rather, *PFMH* was the last of nineteen empirical studies of sexual behavior whose shortcomings were paraded in Kinsey's opening chapter like vanquished enemies in a Roman Emperor's victory parade (*SBHM*, 9). In *SBHM*, Kinsey concluded, contra Terman, that "there seems to be no sound basis for the widespread opinion that the younger generation had become more active in its sociosexual contacts" (397). The bases of these two sexual scientists' disagreement were multiple. First, the kinds of people who participated in the two studies were very different from each other. Terman's study had been recruited through the first marriage guidance center in the country, the American Institute for Family Relations (AIFR). The AIFR was the brainchild of the Californian eugenic activist Paul Popenoe, who had opened it in 1930 to "bring the resources of modern science to the promotion of successful marriage and family life."[5] Kinsey critiqued the bias of Terman's middle-class sample;[6] his own interviewees included bootleggers, male prostitutes, ne'er-do-wells, pimps, thieves, and hold-up men (*SBHM*, 39), precisely the sort of men that many eugenicists would have liked to purify out of the national stock. But sampling was not the only reason for the different conclusions; the patterns of data were not always exactly as authors presented them. Terman's second criticism of Kinsey directed readers of *Psychological Bulletin* to look again at Kinsey's tables where younger men engaged in sexual activity more than older men did, and to particularly note that this trend occurred primarily among the less-educated subgroups (Terman 1948, 453–54). Terman suggested that Kinsey's class-inclusive sample told a story—not of intergenerational stability—but of uneducated men who were dissolving the norm that legitimated sex through the bonds of matrimony.

In contrast to these disagreements about *male* sexuality, Terman and Kinsey found more common ground in their discussions of married women. In *SBHF*, Kinsey adopted a much more conciliatory tone towards past sex research, including Terman's marital study. Therein, Kinsey tended to cite past research when consistent with his own conclusions rather than present his surveys as scientific developments on past errors. In addition to heightened scrutiny of his sampling methods, Kinsey had been chastened by Pomeroy, Martin, and Gebhard, who had compared the results of incarcerated and non-incarcerated women, found considerable differences between them, and insisted that the sample for *SBHF* include only the non-incarcerated subjects.[7] As a result, the sample in *SBHF* was much more middle-class in its composition than the sample reported in *SBHM*. In *SBHF*, Kinsey and his colleagues cited Terman's research to support their claims about the frequency of premarital petting (234), particularly among early maturing adolescents (246), the commonness (352)—but brevity (364)—of marital sex; and the greater desire for marital sex among older rather than younger wives (354). They even conceded the point that patterns of sexual behavior had changed over historical time. About half of the women interviewed had reported premarital coitus, but they had disproportionately been born after 1900 (298–302). Privately, Kinsey adopted a more conciliatory attitude also. When Terman sent a reprint of his 1951 paper on the lack of a relationship between gifted women's orgasms and couples' marital happiness within the gifted cohort, Kinsey replied that the findings were consistent with his own.[8]

However, these two men did not agree about everything. Historian Julia Ericksen rightly characterized Kinsey as a pro-marriage sexual libertarian (1999, 49), but Terman's attitudes both to marriage and to sexual license were ambivalent at best. *SBHF* allowed that "intrinsic capacity" for orgasm might vary in the human female (377), and that wives' orgasms were not the sine qua non of happy marriages (371). But the book also asserted that "the female's failure to respond to orgasm in her sexual relationships is, nonetheless, one of the most frequent sources of dissatisfaction in marriage" (358). For moments such as these, *SBHF* was later remembered by second-wave feminists as one of the rare works by a male author in the postwar period that took women's sexual pleasure seriously.[9] As Jessamyn Neuhaus concluded from the marital advice that Kinsey collected from this period, writings from the postwar period tended to explain variation in married women's sexual satisfaction with respect to women's own personalities and bodies. However, the literature of the 1920s tended to give greater importance to the husband's sexual skill; even Paul Popenoe's writings fit the trend.[10] Rather than locate orgasmic potential in particular women's constitutions, Kinsey concluded that married women *became* more orgasmic over time, and both premarital sex and petting were associated with orgasmic frequency in marriage in his data (*SBHF*, 265, 385, 401). Terman's *PFMH* had adopted a much less optimistic attitude toward the development of married women's pleasures over time.

In this chapter I focus on the conditions under which Terman's book was written, the details of its account of marital happiness, and the effects of its conclusions upon interpretations of Terman's own marriage. I am returning to *PFMH* not to adjudicate the questions opened up between Terman and Kinsey about the determinants of marital happiness in general, or the shifts in sexual habits of particular Americans in the early twentieth century. Rather, I want to extend the sociologists Woolgar and Pawluch's use of the metaphor of "ontological gerrymandering" to draw attention to ways of "making problematic the truth status of certain states of affairs selected for analysis and explanation, while backgrounding or minimizing the possibility that the same problems apply to assumptions upon which the analysis depends" (1985, 216). Within historical writing, sex surveys get gerrymandered as archeological evidence of demographic trends, and as textual evidence of the cultural production of "truth" about sexuality. In this chapter, I want to describe the forms of gerrymandering that have occurred within and about Terman's book, and their lingering effects on the way that we historicize particular marriages and "marriage" more generally.

To date, matters between Terman and Kinsey have been gerrymandered in Kinsey's favor to a dubious extent. The tendency in recent decades has been to credit Kinsey with discovery of the changes in demographics and to erase Terman's earlier pronouncements about them altogether. In *SBHF*, Kinsey conceded Terman's point that rates of premarital sex had been on the rise. When he did so, Kinsey concluded that gender roles and the sexual culture in that decade had been shaped by the writings of Havelock Ellis and Sigmund Freud, movements for women's emancipation, European tours of army duty, increased knowledge of contraception, the anonymity of growing American cities, and a counterproductive drive against prostitutes during World War I (*SBHF*, 299–300). More recent social histories of this period have relied repeatedly on his statistics in their analysis of female sexuality in the 1920s and have made similar claims about these shifts.[11] Such social history has tended to overlook Terman's *PFMH*, and Terman's assertion that Kinsey had gotten things wrong in *SBHM*. Woolgar and Pawluch rightly insist that there can be no account which does not ontologically gerrymander in some manner or another. Accordingly, in revisiting Terman's work, I am "foregrounding" a work that has been obscured by social historians' "backgrounding" of assumptions about the validity of Kinsey's sex surveys.

Revisiting Terman's book might also inform Foucaultian analyses of the history of sexuality, within which married women are doubly positioned. On the one hand, Foucault described the disciplinary power that adheres to "sexuality" in the modern West as historically particular for "a centrifugal movement with respect to heterosexual monogamy" characterized by discretion about the behavior of married couples and a focus on practices deemed to be "unnatural" (1978, 38–39). On the other, he considered a "hystericization of women's bodies" which was carried out in the name of the responsibility they owed to the health of their children, the solidity of the family institution, and the safeguarding of society as one of "four great lines of attack along which the politics of sex advanced for two centuries" (146–47). Were married women's sex lives the point from which normativity departed or the territory over which it advanced? Terman's work was located in time between the 1920s and the postwar period. Neuhaus described these two periods to anchor her description of a shift away from advice manuals' descriptions of obliging husbands who led their wives to pleasure toward those manuals' explanation of variability in that pleasure on the basis of wives' bodies and temperaments. I argue that Terman accomplished such a shift by discrediting Galtonian understandings of marital happiness oriented around ideals and by substituting Queteletian understandings that foregrounded the distinction between "adequate" and "inadequate" marriages. *PFMH* may still have relevance for historians of psychology, and not only because Terman's study is occasionally cited by some twenty-first-century psychologists as pioneering.[12] In the final part of this chapter I will argue that this construction of the intelligent husband as immune to blame for a less-than-ideal marriage is also in the background of biographical writings about the marriage of Lewis and Anna Terman.

The "Chaos of Opinion" about Marital Happiness

The impetus for Terman's marriage study came, in part, from an unhappily married Ph.D. student named E. Lowell Kelly.[13] Kelly was troubled by his reading of the marital advice literature available to him in the late 1920s, and particularly bothered by two recent sex surveys by Katherine Davis (1929) and George V. Hamilton (1929). At one of Terman's Monday night seminars around 1929, Kelly proposed a Ph.D. project on marital happiness, in part for personal reasons:

> Since I was at that time in the throes of a not too happy first marriage, I was much intrigued with these books but differed with their authors with respect to cause and effect. In brief, I took the position that marital compatibility was primarily a function of the personalities of husband and wife and that sexual maladjustment was a symptom of personality incompatibility rather than the primary cause of marital maladjustment.[14]

Terman quickly dissuaded Kelly from pursuing this Ph.D. topic, fearing that Kelly might never finish the work. The following year, Kelly completed an experimental Ph.D. under the supervision of psychologist Walter Miles.

The two sex surveys that had irked Kelly were part of what Terman would later call a "chaos of opinion on the determiners of marital happiness" (*PFMH*, 6). Expert advice on marriage had proliferated in the 1920s as Victorian models of marriage eroded. Sometimes this advice engaged explicitly with the concurrent interest in the new psychological category of "intelligence." IQ researcher Leta Hollingworth's arguments (1916) against pronatalism anticipated Margaret Sanger's distinction (1919) between "birth controllers" and "eugenicists" on the grounds that only the former group considered a woman's duty to the state to be actualized in her duty to herself. Many eugenicists resisted Margaret Sanger's arguments ([1921] 2007) to make birth control the entering wedge of the eugenic movement into the working classes, a view that she promoted after publication of the army IQ tests, and in a context in which scientific racism had galvanized the eugenic movement. Sanger focused on *negative eugenics* to win allies among eugenicists. In her autobiography she recounted rebuking eugenicists for their emphasis on positive eugenics over negative eugenics by making explicit an irony in the sexual politics of these smart men. "'There you are' I said 'a super-intelligent group, the very type for whom you advocate more children, yet you yourselves won't practice what you preach'" (1938, 375).

Californian Paul Popenoe was one eugenic activist who was not receptive to Sanger's rhetoric, considering her supporters "a lot of sob sisters, grandstand players, and anarchists."[15] California's sterilization law of 1909, the third to be introduced in the country, had allowed the involuntary sterilization of "moral degenerates" and "sexual perverts showing hereditary disease." Those convicted of two sex crimes, or three other kinds of crimes, could be sterilized against their will.[16] California denied more of its residents their reproductive freedoms than any other state, and sterilization rates increased dramatically in the 1920s, particularly among second-generation immigrants, and often with familial consent.[17] In 1926, Popenoe, along with the citrus farmer Samuel Gosney, founded the Human Betterment Foundation to research the effects of the state's eugenic policy. Both Lewis Terman and Stanford president David Starr Jordan joined. Gosney and Popenoe's influential volume on California's use of state power to enact eugenic principles inspired and justified the Nazi regime's commitment to eugenics after 1933. Popenoe remained enamored of Hitler's regime throughout the 1930s, even after most American scientists who had turned away from eugenic ideals.[18]

Terman did not renew his membership of the American Eugenic Society in 1935, but he continued to cite the work of Popenoe's Human Betterment Foundation, even after Popenoe left the foundation in 1937.[19] Although Terman collaborated with the AIFR to gather his sample for *PFMH*, he also criticized the Nazi policy of rewarding German women for bearing sons rather than daughters as part of the "chaos of opinion" about marriage that *PFMH* hoped to push beyond (*PFMH*, 49). Popenoe's AIFR had multiple connections to the Stanford psychology department. Three of the AIFR's consultants were former Terman students: Jessie Chase Fenton, Barbara Burks, and J. Harold Williams. In addition to providing the bulk of the data for Terman's marriage book, the AIFR used Terman's psychometric tests to guide couples toward eugenically correct marital decisions, including Terman and Miles's test of masculinity-femininity.[20] Decades later, Terman described how the AIFR also provided "an ideal way to get our data, particularly as it could be collected under the auspices of the Institute instead of under Stanford's name."[21]

Popenoe's eugenic marital-advice writings of the mid-1920s advanced the importance of sexual satisfaction in marriage, at least for his White middle-class audience.[22] Even California's sterilization policy aimed to maintain female sexual pleasure.[23] Popenoe described it as *intelligent* for sex to be a source of power and satisfaction rather than a brute means of bringing offspring into the world:[24] early monogamous marriage—preferably leading to four children—was to be preferred over free love, trial marriage, and egalitarian feminist "50/50" marriages in which women earned their own money.

In this scheme, men bore the responsibility for the couple's sex life. During sex "to go just far enough and not too far" called for good management and constant care on the part of the man (1925, 144). "Prolongation of coitus is much more welcomed by the wife than is vehement haste. The average duration among white people is supposed to be something like ten minutes. Probably it ought to be several times as long as this" (1925, 150).

The early sex researchers whose work had so irked E. Lowell Kelly also linked sex and intelligence. Katherine B. Davis and George V. Hamilton were publicly supported by the American Social Hygiene Association but not by the Committee for Research for Problems of Sex (CRPS).[25] Both used maturity and intelligence as rhetorical grounds for normalizing the people who answered their probing questions. Among the 2,200 "normal women" surveyed by Davis were married participants "of sufficient intelligence and education to understand and answer in writing a rather exhaustive set of questions as to sex experience" (Davis 1929, xi) and single women who were college graduates of at least five years who "would be more likely to recognize the need of such a study as that proposed than would younger women." Hamilton conducted interviews with one hundred New York married couples who were "serious-minded, more or less importantly occupied and well above the average as to intelligence and cultural attainment" (1929, vi). The couples' responses showed how two very different phenomenological worlds could exist in the same bed. While men thought that foreplay and "prolonging the act" would bring their wives to orgasm, wives mentioned tiredness, "a happy evening together," and "his fondling her, making love to her" as more important (172–73). Only half as many husbands as wives claimed that the wife never had an orgasm (171). In spite of these differences, or perhaps because of them, Hamilton praised the "intelligence and honesty" of his participants for speaking to him so frankly (190).

The most popular book of marital advice to emerge from this period was Dutch physician Theo Van de Velde's *Ideal Marriage: Its Physiology and Technique* (1933).[26] Like Popenoe's writings, Van de Velde's book described husbands as intelligent if they educated their wives in the experience of sexual pleasure. Van de Velde described a husband's "neglect of love-play" as a "sin of omission" that was "unpardonably stupid" (1933, 148). Unlike the speedy associations that characterized the natural brightness of Terman's gifted children, Van de Velde equated male sexual smarts with slowness. Women took longer to reach orgasm than men did, and husbands should slow down and notice the physical signs that coitus could proceed:

Is not the necessary excitement and desire on her part, to cause swelling of the labia, dilation of the vulva, and erection of the clitoris—then, as these manifestations are normal and desirable *before coitus, it is both stupid and grossly selfish of the husband to attempt it, if they are absent.* (1933, 167)

Van de Velde frequently linked sex and the intellect through the metaphor of musical performance. Foreplay was a "prelude" to the main event (151), the wife's orgasm was "concert pitch," and the chapters of the book were even punctuated by sections called "intermezzos," which consisted of aphorisms about sexual pleasure and knowledge. The musical analogy seemed to explain the primitive roots of cultural activities (148), and also explained why sex had a complicated relationship to speed: "We realize this general rule of aesthetics in music, where it is most effectively used; in the drama, where a short interval enhances interest in the action when it recommences; and in the erotic art through coquetry" (149). However, the most vivid musical analogy in *Ideal Marriage* afforded only one gender the position of the sexual genius:

Balzac has proclaimed that "In love—quite apart from the psychic element— woman is a harp who only yields her secrets of melody to the master who knows how to handle her." But who can play this delicate human harp aright, unless he knows all her chords, and all the tones and semitones of feeling? Only the genius—after long practice and many discords and mistakes! (242)

In short, the ideals in Van de Velde's account of "ideal marriage" obliged husbands to manage the timing of sex to draw out their wives' precarious but considerable capacities for sexual enjoyment. "Intelligence" was not the dominant way of thinking about these differences between husbands, nor was the "intelligence" and "stupidity" mentioned in these accounts the same thing as the "Intelligence" and "Stupidity" ontologized in Terman's science. Rather, much as Kinsey's claims about the personalities of adolescent boys lacked the rigor of psychological testing, so Terman might well have been irked by the citation of ideas about male intelligence in these works, particularly as that intelligence was embodied rather than associative, characterized by slowness and rhythm rather than speed, and oriented towards deferring gratitude in the self with the goal of increasing a woman's pleasure rather than solving abstract problems. How might intelligent men at Stanford work out their frustrations with a literature that looked like this?

Gerrymandering Less Than Ideal Husbands: Psychology

E. Lowell Kelly's hypothesis that *personality* variables might predict marital happiness was timely. As Ian Nicholson notes in his biography of the psychologist polymath Gordon Allport, "in 1920s America, personality was the language of mediation. Combining past and present, masculine and feminine, sacred and secular, the category's vast ambiguity provided welcome shelter for the uncertainties for the new and self-consciously modern age."[27] Indeed Chicago sociologist Ernest Burgess, whose work Terman would cite favorably in *PFMH*, was also pursuing Kelly's hypothesis that personality was the mediator of marital success and failure.[28] Terman successfully secured CRPS funding for himself and Kelly to conduct complimentary research projects on marital happiness; Terman was to do a cross-sectional study and Kelly to begin his longitudinal work.[29] In crafting these plans in the early years of the Depression, Terman had Kelly's future career in mind as much as Depression-era rising divorce rates. Much to Terman's surprise, Kelly was hired to the faculty of the University of Connecticut in 1933, and Terman was left needing collaborators to complete the marriage research. In Connecticut, Kelly began his longitudinal study by contacting engaged couples through newspaper advertisements, as he had originally intended.[30]

Left without Kelly's assistance, Terman turned to Popenoe's AIFR for help. Between December 1934 and April 1935, Donald Wilson, the Research Director of the AIFR, collected much of the data included in Terman's volume. As with the research assistants on the giftedness project, Terman underestimated the effort and smarts needed to construct the samples, later describing Wilson as a man "with a lot of ability, but rather scatter-brained and jumps from one thing to another."[31] Terman's survey hardly made it easier for Wilson to focus on a single task. Initially, Terman wanted 1,000 couples who would complete the survey simultaneously but independently. The survey was to include a pencil-and-paper IQ test, and Terman particularly wanted *unhappy* couples to be sampled. Terman and Popenoe also disagreed about which tests to include, and Terman changed his mind as to what to include based on his rereading of Hamilton's book on marriage. Wilson's task was complicated further by the reluctance of some people to take the time to complete surveys when they were short of food because of the Depression.[32] In addition to the Institute's clients, Wilson ultimately worked with a wide range of groups—the California Eugenic Society, Los Angeles churches, and labor unions among them—to construct the ultimate sample of 792 couples. Like other sex researchers, Wilson emphasized the *intelligence* that participants in the research would show by donating their time—and trusting their privacy—to the project. Figure 1 shows an recruitment advertisement extracted from a program for the Second Annual All-day Harbor Regional Conference on Family Relations at the YWCA in Long Beach California on December 8, 1934. The advertisement is among Terman's papers. Couples who took part in the study at this event were reminded that the study's success "depends on the cooperation of intelligent married people who want to do something to promote successful marriage for others in the future."

When researchers use questionnaires to measure marital happiness, they ask a lot of the "intelligence" of research participants that questionnaires are capable of drawing out. *PFMH* is a remarkable book within Terman's oeuvre because of his awareness that the measurement tool does not reach the psychological experience that it targets. Terman reminded his readers that "it must be admitted at once that the quantification of such imponderables is but a convenient device" that is "rough as well as arbitrary" (*PFMH*, 4). Moreover, the definition of marital happiness was defined as subjective; only happiness "as experienced by our subjects themselves" can be measured, not happiness *itself* (49). Notably, this empirical project also stopped short

---

**UNIQUE OPPORTUNITY TO TAKE PART
IN STUDY OF MARITAL HAPPINESS**

The Institute of Family Relations has undertaken to study 1,000 normal married couples, in order to get more scientific data on the factors that make for happiness. Husbands and wives are asked to fill out anonymous and wholly secret questionaires. This will require 30 to 45 minutes. Participants will receive a copy of the report when completed, and also one on 400 couples already studied. Husbands and wives must come together for this purpose, and meet in the main auditorium at 5 p. m. or 9:30 p. m., as may be most convenient for them. ..It is not necessary to be a registered member of the conference, in order to take part in this study. ..Its success depends on the cooperation of intelligent married people who want to do something to promote successful marriage for others in the future. Will you help?

D. P. Wilson, Research Director.

For further information concerning its work, address
THE INSTITUTE OF FAMILY RELATIONS
331-3 Consolidated Bldg., 607 S. Hill St.
Los Angeles, California.

Fig. 1. Recruitment Advertisement for Terman's Marital Happiness Study.

of questions about "autoerotic practices or indulgences in extramarital intercourse" (37) in spite of Wilson's advice that an early wave of participants called for *more* questions on these matters.[33] Here as elsewhere, Terman insisted that men could not be asked to recall when they reached puberty: "information on this point is difficult to secure for males because subjects are not able to date the attainment of puberty with anything like accuracy" (*PFMH*, 253).[34]

Because self-report questionnaires were gathered separately from wives and husbands, Terman could gerrymander very different accounts of the couple's most intimate moments. Husbands gave slightly longer estimates than their wives as to how long intercourse lasted (*PFMH*, 296). Terman noted that to resolve differences, "records in sufficient quantity based on chronometer readings are hardly possible, though chronometry experiments had shown that men's judgments of time were more reliable than women's" (294). However, he did not simply trust husbands to be more reliable witnesses than wives to the timing of intercourse, but rather undercut the claims of those particular husbands who judged that they embodied the kinds of ideals described by Popenoe and Van de Velde. Among the husbands who reported that they could prolong the duration of intercourse, examination of their wives' questionnaires showed that only one-third reported that intercourse typically lasted longer than ten minutes (300). Terman highlighted these wives' accounts to position the patient, rhythmic, unselfish, and intelligent husband of an earlier literature as an impossible ideal.

Indeed, for Terman, there was little that was "ideal" about marital sex at all; it was a "lily" that "has already been sufficiently gilded" by previous expert authors (247). Critical of free love, companionate marriage, and other recent experiments in living, Terman never endowed marriage with the kind of Christian or Platonic sanctity that might make marital sex sacred or transcendent either. Terman's questionnaire asked wives if they had experienced "a climax of intense feeling followed by quietude and a feeling of relief" (300).[35] Contrary to Kelly's original hypothesis, this experience was correlated with "the responses of each spouse to every one of the individual items which enter into the happiness score" (303). Wives' orgasms also made a unique statistical contribution to marital happiness (305), and received more weight in the measurement of marital happiness than any other aspect of the couple's sex life (304). Clearly there was some sort of lily here, but to salvage the account of its previous over-gilding, Terman constructed the chaos of past opinion on marriage as uniformly prioritizing the wife's orgasm while "our data do not support the frequently expressed opinion that [lack of orgasm] is the one major cause of unhappiness in marriage" (304–5). As Henry Minton has noted, Terman's description of the career ambitions of wives subtly urged women to lower their expectations lest they end up unhappily married.[36] Married women were advised by this book to lower their expectations in the bedroom also.

In Terman's marriage book, marriages differed less in the extent to which they approximated something like Van de Velde's ideal and more in terms of whether those marriages were "adequate" or "inadequate." Married women who "never" or "sometimes" experienced orgasm while having sex with their husbands were described as "inadequates," and a long appendix to *PFMH* compared "adequate" and "inadequate" wives' responses to various parts of the questionnaire. Here, Terman attempted to unravel the "puzzling mystery" of "the inability of a large proportion of women to achieve the climax." Finding little grounds for difference in "background experiences" or other sexual variables, Terman concluded instead that deep-rooted *personality* differences were in question. Inadequate wives had indecisive personalities, were more conformist, preferred "audience" to "player" roles, were more romantic and aesthetic, and were given to extravagance, particularly when it came to spending money on clothes (402–3). Terman alighted on the explanation that the "constitutional factors" that caused orgasm inadequacy were also involved in "neurasthenic tendencies, diminished responsiveness, and lack of zest, vigor or colorfulness of personality" (375). Admitting that the reason "why one woman out of three fails to achieve orgasm, or achieves it only rarely, is still a mystery" (407), Terman nonetheless located the cause of orgasm adequacy *within* women themselves rather than in their marital relationships or in their husbands' abilities. By attributing these differences in women's sexual experience to imagined constitutional differences, Terman made any causal relationships between variation in husbands' intelligence and variation in their wives' satisfaction more obscure.

However, "intelligence" remained relevant in this work, even if husbands' performance in the bedroom was constructed as unrelated to it. Couples were informed that the questionnaire was *not* an Intelligence test, but they were still asked to evaluate their own and each other's "mental ability." Wives were more likely to be happy if their husbands were of similar or higher ability than themselves. In contrast, husbands were happiest when they considered their wives neither smarter nor more stupid than themselves (192). One connection between male intelligence and good marital sex remained;

husbands who thought that sex *was* important to marriage were *less* likely to have happy marriages (165–66). Terman had effectively abnormalized the earlier idealization of married sex as both nonexistent and harmful. Well-matched clever people had *adequate* sex and did not strive after impossible ideals.

It would be possible to gerrymander these data differently and to bring to the foreground some matters that would argue *against* Terman's conclusions. In his search to unravel the puzzle of orgasm inadequacy, Terman did examine differences between husbands, and reported that the husbands of "inadequate" women were younger, less educated, more religious, more prone to conflict with their fathers, were later to learn where babies came from, had parents who were more reticent to answer their questions about sex, and were more likely to have learned about sex from other children (403). These differences might suggest that these men—if taken at their word—had experienced sexual educations that bore an uncanny resemblance to those experiences that the 1920s authors of the "chaos of opinion" feared the most. Terman did not consider the possibility that the husbands of non-orgasmic wives were sexually immature, lacked an educated intelligence of sex, were "stupid" in Van de Velde's terms, or were "brutish" in Popenoe's. Rather, he psychologized these empirical differences as evidence of *personality* factors. The husbands of inadequate women preferred to be with others when they receive bad news or were under stress, disliked masculine women and feminine men, accepted the Bible uncritically, and thought that religion did more good than harm.[37] Terman concluded that the husbands were "somewhat lacking in positive and dynamic traits of personality" and that "like tends to mate with like" (403–4).

Terman's book was a popular success, but its claims did not go uncontested within academic psychology. In a review in *Psychological Bulletin* that foreshadowed Terman's review of Kinsey's in its cautions about self-report data, psychologist Harry Hollingworth pointed out that Terman's conclusions about what-caused-what were underdetermined by his correlational data, particularly as it was unclear where halo effects had or had not occurred. Hollingworth feared the book was so poor that it would bring personality psychology into disrepute:

> If this volume ever gets into the hands of the newspaper boys they will get a lot of fun out of the solemn discovery that it is unfavorable for happiness for spouses to disagree on their liking for Pershing, that it is ominous for both mates to feel the same way about crossing the street to avoid meeting some-

one, and that among the best differentiating items of happy from unhappy spouses are mutual interest in dental work and joint fondness for men who use perfume. Thus low have the personality traits fallen. (1939, 192)

Terman's response (1939) to Hollingworth introduced the first of E. Lowell Kelly's publications (1939) on his longitudinal study; a two-page report to the effect that knowledge of personality and "background" factors prior to marriage predicted marital happiness two years later. However precocious the predictions of Terman's marriage research might have been, his career advice to Kelly was certainly predictive. It would be another eleven years before Kelly would publish another paper (1950) about this data, and another forty-eight years before the next report (Kelly and Conley 1987) would appear—the first to report on the relationship of personality to marital happiness.

This closing-down of questions about married men's variation from the ideal provides a useful vantage point to consider the relationship between gender, sexuality, and intelligence in Terman's thinking. The adjustment of the gifted children, after they had matured, became linked with this argument about the determinants of marital happiness. Between 1939 and 1941, Terman received $7,300 in funding from the CRPS to study the marital adjustments of his gifted cohort and the control subjects in adulthood, and the findings were reported in *Genetic Studies of Genius*, vol. 4 (1947), co-authored with psychologist Melita Oden.[38] Therein, Terman and Oden could confidently announce that the gifted formed marriages that were slightly happier than those reported by the couples studied for *PFMH*, and that differed neither from the control subjects or the *PFMH* couples on any of the items assessing "sex adjustment" (239–63).

In other areas of his research in the late 1930s, Terman emphasized that the important variability among men was in the disembodied area of intelligence. While Olga McNemar was assisting Terman on his marriage study, her husband collaborated with him to reinvigorate the "variability hypothesis," which specified that men's intelligence was more variable than women's intelligence.[39] Stephanie Shields's careful history (1982) of this hypothesis shows how it had, in earlier decades, divided Intelligence psychologists along lines of gender, and how the hypothesis had collapsed into technical controversy before McNemar and Terman revived it. Thus in Terman's writings with the McNemars, differences among men that matter get dislocated out of the embodied practices of heterosex and firmly ontologized in the field of intellectual differences.

Gerrymandering Less Than Ideal Husbands: Biography

Gerrymandering occurs also when historians write about psychologists, just as when psychologists write about their research subjects. Both forms of accounting require foregrounding and backgrounding. Like the psychologists who interpret self-report data in different ways, historians gerrymander the past when we trust some people as witnesses and make others' lives into events that others have reliably witnessed. Terman was right that Kelly would make little from his prospective study. However, his longitudinal data was later exploited by historian Elaine Tyler May in her history of American marriage, *Homeward Bound* (1988). While May's history deserves the praise it has garnered, I want to draw its gerrymandering of the relationship between history and the human sciences into relief here. Like other social historians, May relied on Kinsey for evidence of the demographic shift in premarital sex after World War I. She also used the open-ended questionnaire responses and letters that couples wrote to E. Lowell Kelly to individualize the narratives created by Kinsey's statistics. Showing the typical preferences of a historian over a psychologist, May favored letters over questionnaires as less constrained ways of speaking from the heart about marital experiences: "In these open-ended responses [to Kelly's questionnaires], freed from Kelly's categories and concerns, they [the middle-class White married couples] poured out their stories" (1988, 12). May could have written a different account less invested in disciplinary boundaries between social history and social science. Indeed, the very existence of the open-ended items could have suggested that Kelly knew that his own categories and concerns might not have been ideal. Moreover, Kelly's own marital history disappears from this history. One would not know from May's account that Kelly was married twice, participated in the historic transformations of marriage that she describes,[40] that he was Terman's protégé, or that his interests in marriage inspired a major and impactful book on the subject (more consistent with May's conclusions about demographic shifts than the volume by Kinsey that she cites). In other words, May's use of data shows how some human science gets backgrounded as historical fact and other human science gets foregrounded as "social construction." In this final section of the chapter, I want to foreground the figures of the blameless less-than-ideal husband and the inadequate wife, to consider how they characterize the gerrymandering in the *biographical* accounts of Anna and Lewis Terman's own marriage.

Before continuing with that task, it is important to note that Terman so gerrymandered when he contributed to the biographical literature of another male intellectual. Thorstein Veblen was, arguably, the most important American economist of his generation, and accounts of his marriage with his first wife, Ellen, are varied. These two married in 1888 and divorced in 1911, while living at Stanford. The divorce happened apropos of Thorstein's affairs with younger women, which also occasioned his dismissal from the university. Upon his departure, Ellen became intimately involved with divorcée Clarissa Dixon and a patron of her son Henry Cowell. When Ellen Veblen died in May 1926, her body was given to research. Her pathologist A. W. Meyer described her corpse to Lewis Terman, who later wrote to Joseph Dorfman, who was later Thorstein Veblen's biographer—that Ellen had an "infantile sexual development" and was hence incapable of "having had, or given to her husband, a normal sexual life." Terman wrote to Dorfman with the aim of reducing the blame attributed to Thorstein for his affairs and for the end of their marriage.[41] Dorfman's biography of Thorstein noted that "examination showed her [Ellen's] physical development to be physically retarded, scarcely even to have passed beyond early adolescence." Dorfman did not repeat the inference about the Veblen's married life that Terman drew from this physical description of Ellen Veblen's body.[42] However, later authors who have written about Thorstein Veblen have done precisely that.[43]

In contrast to his contribution to the history of the Veblens' unhappy marriage, Terman told us little about the relationship between his own marital experiences and his intellectual passions. However, in his precociously authored contribution to Murchison's *History of Psychology in Autobiography*, he credited his wife with remembering his own early account of the relationship between family life and scientific interests:

> In 1899, a year after I left Central Normal College, I was married to Anna B. Minton, a teacher whom I had met at Danville three years earlier. She had begun teaching at even an earlier age than I, had come under the influence of the same teachers at Danville, and had exactly the same objectives for me as were already shaping themselves in my own mind. Less than a year later, our first child was born, bringing a new and vast psychological interest into my life. My wife reminds me (I had forgotten it) that two to three years later, while a student at Indiana University, I told her my interest in our baby had determined me to become a psychologist. (Terman 1930, 309)

The Termans did not remain the happy young couple of this account. Anna and Lewis's son Fred recalled much later to May Seagoe that the family holidayed together only once after coming to Stanford.[44] By the time that Lewis

Terman's autobiographical chapter was written, he had had a series of affairs with younger women. Writing Terman's biography, May Seagoe was vexed as to how to account for these happenings, and to her credit she wrestled with the problem. She shared with Terman's surviving friend, E. G. Boring, her impression of Terman as a "fundamentally shy man whose attempts to get close to bright and warm individuals might well include sexual overtures where women were concerned," and she conjectured that the affairs "were probably always secondary, both to his work and to his sense of obligation to the family."[45] To Harry Harlow—Terman's former graduate student—Seagoe described Terman as a "man of ability and high work-dedication" straining not only under an early marriage but also "a pull toward those who can share ideas as well as social companionship."[46] Harlow was no stranger to empirical studies of love or marital instability,[47] and he replied that Seagoe was writing with "total insight and understanding"; Terman's affairs "represented intellectual and social ties more than anything else."[48]

Seagoe's biography ultimately described "an affair with a very attractive young woman one summer, as well as two or three other liaisons" and the departmental gossip about "the cabin in the hills, the outside entrance to his study at home, and his plan to build a new study in a Los Altos golf tract, where he would be pretty much secluded" (1975, 147–48). Seagoe invoked several explanations of the affairs with a sense of hurried panic, as if none of them quite provide the secure footing she needs. In less than one page of text (131), she attributed the affairs to Anna's intellectual limits ("Anna was an exceptionally supportive wife, but she had never pretended to share his intellectual interests"), Lewis's strong masculine interest in sex ("learned from his early farm years among neighbor boys for whom it was the prime quality of life"), nature ("sexual intercourse was a natural activity of man, and in a rich life it was to be promoted, provided harmful effects were avoided"), the acceptability of extramarital sex in 1920s Stanford (where "certainly extramarital affairs were common"), and the dearth of women on the Stanford campus (such that "sexual competition was exaggerated out of all proportion"). On that campus Terman nevertheless "found mature women who were intellectually stimulating" to whom he was particularly attracted. Seagoe seems to hop from each to each as they crumble into implausibility, alighting ultimately on the conclusion that "whatever the answer, Terman's professional productivity did not suffer" (132).

Thirteen years later Henry Minton's somewhat more critical biography of Terman poignantly noted that Seagoe had unintentionally revealed the father's affairs to the once-cherished infant son. Unlike his sister Helen, Fred Terman had no knowledge of his father's affairs prior to Seagoe's writing about them (Minton 1988, 164). Rather than gloss Terman's affairs as intellectual passions, Minton constructed a biographical relationship between those affairs and Terman's later psychological conclusion that "sexual compatibility was not closely related to marital happiness." For Minton, such a conclusion "would provide a means of salving his [Lewis Terman's] conscience" about the earlier affairs (166). Minton's biography does not say what aspect of the affairs burned particularly in Terman's conscience. Curiously, at this point in his text, Minton referenced a page in the appendix of *PFMH* that does not mention extramarital affairs at all. On that page, Terman noted that "the influence of sexual factors [on marital happiness] is at most no greater than that of the combined personality and background factors, and . . . it is probably less," and that "if the [wife's] orgasm is not established within the first year of marriage it is unlikely ever to be" (*PFMH*, 376).

If Minton thought that Lewis felt guilty about the state of his marriage or Anna's experiences in their bedroom, then he did not say so directly. But the discussion of Terman's affairs in Minton's biography has a family resemblance with Terman's account of Ellen Veblen's inability to please her husband and with the attribution of "orgasm inadequacy" to the underzealous personalities of wives in *PFMH*. At various points, Minton describes Anna Terman during this period of their marriage as domineering, suffering from empty nest syndrome, and as someone who became increasingly controlling of her career-conscious husband after their children were grown (1988, 166). In other words, biographical writing about the Termans is largely consistent with the scheme he laid out in *PFMH* in which husbands are assessed by their intelligence, wives by their character, martial happiness is a consequence of variation in the latter and not in the former, and marital heterosex disappears from view.

The *heteronormativity* inherent in failing to draw out more elaborate, psychologically penetrating, character-destroying inferences from the heterosexual affairs of such smart men is also consistent with the logic of *PFMH*.[49] Therein Terman dismisses Freud's Oedipal theories linking parental attachment to heterosexual marital happiness but assumes a reality to the Oedipal explanation of latent and overt homosexuality, as common opinion, psychoanalytic theory, and his own empirical research with Catharine Cox Miles all suggested "the 'mamma's boy' as a particularly poor marriage risk" (216). Consistent with this "foregrounding" of the explanation of homosexuality rather than heterosexuality, the *homosexual* affairs of the married Alfred Kinsey have incited explanation of various forms, including the psychoanalytic. Both Jones and Gathorne-Hardy described marital sex as initially painful and unsatisfying for the young Kinseys. The matter was resolved by a gynecological examination and a straightforward operation on Clara's genitals.[50] James Jones narrated the subsequent married life of Clara and Alfred Kinsey as purged of eroticism, love, and honesty (to the point of describing Clara as a woman who *might* have been attractive, had she worked on her appearance) (1997, 164). The secret shames of Alfred's homosexuality and masochism explain this less-than-ideal marriage. In contrast, Gathorne-Hardy's biography relies on a discourse in which sexual alterity can be a basis for exceptional insight.[51] For Gathorne-Hardy, Kinsey is a happily married bisexual, and bisexuality grants him an "an almost ideal position, one might think, for someone who was studying sexual behaviour in both sexes" (1998, 83).

Jacques Derrida (1995) used the term "archival violence" to describe the ways that historians' ontological commitments undercut those of the human scientists that they write about. In an earlier essay (2005), I suggested that the dynamics of lesbian, gay, and queer historiography might provide a more useful starting point for considering archival violence that biographers enact on the human scientists that they write about than the Oedipal narratives of psychoanalysis at the center of Derrida's account. Both the continued psychoanalytic suspicion leveled against Kinsey, and Terman's argument that the Oedipal hypothesis be used to explain homosexuality, exemplify how heterosexuality has provided a kind of shelter from undoing explanations of a human scientist's less-than-ideal conduct. But Terman's reputation is not beyond reproach simply by virtue of the fact of his heterosexuality; discourses of sexuality and the intellect intersect. Journalist Joel Shurkin's book *Terman's Kids* (1992) exemplifies that genre of tragic writing about gifted children in which the promise of youth fails to make good, and the reader enjoys schadenfreude as the arrogance of the intellect is undone.[52] Shurkin casts doubt on Terman's character apropos of his extramarital affairs by linking them with the giftedness study:

> Without a blink or a shudder of hypocrisy, he spent considerable time studying what made marriages happy (with no regard to the unhappiness of his own) and sex and sexuality (with no apparent introspection of his own). In some ways, it was the single most insidious flaw in the Terman study of the gifted. (1992, 85)

Both Terman and the biographers that he impressed characterize gifted boys and men as paragons of virtue. But as the Galtonian normalization of smart men has been premised on sexual normality from the outset, Terman's personal failings can be foregrounded to trouble that normalization, as in Shurkin's book.

More generally, Terman's writings on marriage, and writings on his own marriage, ought to inform Foucaultian understandings of the history of sexuality and its relationship to normalization. Terman shifted the discourse of marriage away from a normalization with respect to new "ideals" of pleasure, foregrounding the distinction between "adequate" and "inadequate" marital sex. The heterosexual couple had been the silent "centrifuge" about which Queteletian normalization of homosexuality has radiated, but married women have also been rendered hysterical, in Foucault's terms, or "inadequate" in Terman's, in the normalization of heterosexual marital happiness. Thus heteronormativity and androcentrism intersect in backgrounded characterizations of the disembodied intellectual interests of married men. However, such strategies of normalizing men can be reversed, and Shurkin's text will not be the last one we encounter that shows how the normalization of smart men becomes precarious when their sexualities risk notice.

4

Queer Individuals

Their Nature and Nurture

What Terman *Didn't* Say

In the previous two chapters, I examined silences in Terman's thought, silences toward which I was led from his review of Kinsey's *Sexual Behavior in the Human Male*. In this chapter, I examine a silence within the review itself. Terman's review of Kinsey is distinctive for the *lack* of attention paid to Kinsey's finding about the remarkable frequency with which adult men reported homosexual sex. Terman was not generally disinterested in homosexuality; rather, he tended to endorse Oedipal theories of the causes of it. One of his books, *Sex and Personality*, is very much preoccupied with the explanation of homosexuality as a form of gender inversion. I am not the first to attempt to explain why Terman wrote the particular review that he wrote. So I will begin with some existing explanations of the review that don't seem, to me, to convincingly explain this particular silence.

First, Kinsey himself judged that Terman's review was motivated by moral prudery,[1] and Pomeroy's biography of Kinsey (1972, 287–94) categorized Terman's review among the set of moralistic responses to *SBHM*. In a preliminary draft of his review, Terman did describe how Kinsey's research deviated from a Puritanical code; and in the published version of that review, Terman critiqued Kinsey for going beyond a *description* of his data and prescribing sexual morality inappropriately. However, if moral prudery were all that motivated Terman, then he might have expressed horror or disbelief about the frequencies with which Kinsey reported American men to have carnal knowledge of each other. Of course, moralism about sex does not have a direct relationship to the silence/knowledge binary. Henry Minton's biography effectively bypassed the need to explain Terman's silence by interpreting

Terman's reference to "uninhibited sexual activity" as a coded reference to Kinsey's comments on male homosexuality (1988, 232). But Terman clearly felt no need to encode his views about adolescent masturbation, premarital sex, or other matters in his public response to Kinsey. Why would male homosexuality have a particularly close relationship with discretion? As Terman's other writings also mention "sexual inversion" and "homosexuality" explicitly, and other Kinsey critics called attention to these frequency data almost universally, Minton's explanation might seem plausible only because of a modern propensity to associate homosexuality with secrecy.[2] Terman's silence is not readily attributed to moral prudery. Let's move on.

Perhaps the shape of Terman's review can be explained by such motives as professional jealousy and revenge.[3] The impact of Terman's own sexological books, *Sex and Personality* and *Psychological Factors in Marital Happiness*, were vastly overshadowed by the surprising sales of Kinsey's *SBHM*.[4] Terman's comments on masturbation and on pre- and postmarital intercourse had zoned in on passages of Kinsey's work where his own studies had been cited and reinterpreted. Yet, such jealousy and revenge would have led Terman to target Kinsey's claims about homosexuality rather than ignore them. Like the passages on masturbation and marriage, the chapter of *SBHM* on homosexual outlet made a negative example of Terman's work. Specifically, Terman and Miles's book *Sex and Personality* exemplified, for Kinsey, the kind of flawed work that had assumed that "an individual's choice of a sexual partner is closely related to the masculinity or femininity of his personality" (*SBHM*, 637–38). While Terman critiqued Kinsey's use of his work on gifted children and marriage, he chose not to respond to Kinsey's criticism of *Sex and Personality* as exemplifying psychological work that was limited by preconceptions about the relationship between sexuality and gender (637–38). Why?

Perhaps then *I* am making something out of nothing, and Terman was simply disinterested in Kinsey's views about sex between men? Private correspondence between the two suggests otherwise. Terman read Kinsey's 1941 critique of hormonal models of male homosexuality, the work that Kinsey presented at the AAAS where he encountered Robert Yerkes. Terman wrote that he thought Kinsey "doubtless right in the belief that there is no such sharp line of division between this class of people and 'normals' as most people have assumed." Terman urged Kinsey to materialize such a difference: "put your data again through the statistical mill, this time leaving out of account all those subjects whose homosexual contacts were of the milder sort, as for example mutual masturbation and things of that sort."[5] In response, Kinsey described his now famous seven-point continuous scale anchored by exclusively homosexual and heterosexual histories at its lower and upper ends.[6] Terman seemed to miss the distinction between categories and continua, responding that he was "delighted to know . . . that you are already treating your homosexual data to bring out the type of things suggested in my earlier letter."[7]

Terman then asked Kinsey to help his colleague Robert Ross, who was re-searching Terman's test of "Masculinity-Femininity" to measure homosexual tendencies.[8] Reasserting binary logic once again, Terman told Kinsey that "the cases he would especially like to have data on from your group are those at the rather extreme homosexual end of your distribution." Kinsey replied that he would write to Ross, but never did. Ross pursued this project by test-ing Navy recruits with the Masculinity-Femininity test with a view to revis-ing the test so that it could discriminate homosexual and heterosexual men more accurately. Ross never published this work, in spite of a small grant from the CRPS to support its completion.[9] Indeed, when Ross published later studies on male college students, those studies took the Kinsey data as *the* normative estimates of population frequencies of homosexual behav-ior among men.[10] In short, self-censorship, moralism, professional jealousy, disinterest, and the general shape of criticisms of Kinsey's *SBHM* all fail to explain why Terman might *not* have called attention to Kinsey's remarkable claims about the frequency of male homosexual experience.

These failures of cooperation between Kinsey and Terman, and failures of historians to convincingly explain them, motivated me to question whether the "homosexuality" that each man was talking about actually was actually the same thing or not. Eve Kosofsky Sedgwick argued for an understanding of different theories of male homosexuality as varying by their location in an unsettled epistemological space defined by the relationship between homo-sexuality and gender inversion, and the extent to which homosexuality is "an issue of active importance primarily for a small, distinct, relatively fixed ho-mosexual minority" or "an issue of continuing, determinative importance in the lives of people across the spectrum of sexualities" (1990, 1). In this chap-ter, I would like to locate Kinsey and Terman's ships-in-the-night comments about homosexuality in this fractured epistemological space, and to expand the reach of Sedgwick's schema further into the human sciences. In her his-tory of the American sciences of homosexuality, Jennifer Terry (1999, 13) cleverly notices that Lewis Terman and Alfred Kinsey were rare conspecifics who shared the distinction of garnering funding from the Committee for Re-search on Problems of Sex (CRPS) for *behavioral* studies of human sexuality. Terry attributes the unusual faith that the CRPS placed in each man in part to his prior successes in unrelated scientific fields—intelligence testing in Terman's case and insect biology in Kinsey's. To explain Terman's silence, it is necessary to understand how insect biology and intelligence testing afforded very different ways of knowing about male homosexuality. While Sedgwick described the crisis of definition of homo/heterosexuality as leaving its mark on the meaning of many other modern binary definitions,[11] this chapter then aims to use the shape of Terman's review to add two binaries to Sedgwick's list; that between insect/human sociality—an ontological binary most often configured by normalizing analogy, and gifted/average intelligence—an ontological binary most often configured by normalizing comparison.

Queer Individuals and the Western Frontier

One of the queerest things about the Kinsey reports is that they were written by an entomologist. In *SBHM*, Kinsey described his method of sex research as "taxonomic," and taxonomy as "the oldest of the biologic sciences" (17).[12] Unsurprisingly, several biologists have attempted to describe how Kinsey's training as a biologist affected his studies of human sexual behavior. The best essay of this type is the cleverly titled essay "Of Wasps and WASPs" written by the great taxonomist and historian of science, Stephen Jay Gould (1988).[13] Gould judged that Kinsey's sex surveys had followed good taxonomic principles in three principle regards. First, Kinsey had relied on large samples. Kinsey was a copious collector, who had been known as "get-a-million" during Ph.D. days at Harvard, because of his passion for collecting biological specimens. His sex surveys, which reported on thousands of Americans, were unsurpassed in sampling size for decades. Second, Kinsey had searched for biological diversity, sampling in a way that put him in some tension with the CRPS's goals for his work. Third, Kinsey had suspended prejudgment about the character, value, and morality of the forms of life observed. Indeed, Kinsey's description of variation in human sexual behavior with a rhetoric that eschewed moral judgment often led his own science to become a target of moralizing criticism.

However, Gould's own essay contains some prescriptions and normative standards of its own. Gould argued that the taxonomic approach was more suited to gall wasps than to humans, and that Kinsey had failed to recognize that social science was a much more inherently value-driven process than biological science. Yet Gould also divided the taxonomists of the early twentieth century into two categories: "lumpers"—who tended to categorize variant individuals a bit too much within a single species—and "splitters"—who tended to nominate a greater number of species than was strictly necessary. Gould's admiration of Kinsey was tempered by his classification of him as a "splitter" who authored a "bloated taxonomy of full names for transient and minor local variants" (1988, 161). In particular, several of Kinsey's taxonomies assumed that spontaneous mutations had created entirely new species. But Gould insisted that nonidentical American gall wasps kept their "central tie to the rest of the species—the ability to interbreed" (162).[14] Where Kinsey's taxonomy foregrounded continua rather than essentially distinct forms of life, Gould insisted on heterosexual reproduction as the criterion by which life forms could be ordered as essentially separate species. Where Kinsey had claimed Darwinian discoveries about the origins of new American species, Gould saw Kinsey as making a difference out of no difference at all.

While Gould valued Kinsey's rejection of that form of normalization that is little more than a "departure from the standards of the individual who is passing judgment," in the domain of sex, there is an internal speciesist contradiction in Gould's critique. *His* evaluation of Kinsey depends on a classification of the subspecies of *Homo academicus taxonomica* into lumpers, splitters, and rational normative others (a classification that is more value-laden than it first appears). These pejorative terms are well known among taxonomists; Kinsey mentioned them in one of his very first scientific papers, and rhetorically positioned his approach to taxonomy on the normative thin line between lumping and splitting:

> By the employment of varieties I am open to both the charges of being a "lumper" and a "splitter," the former in my treatment of species, the latter in my recognition of varieties; but in any event the scheme is employed with consistency, and does portray the different degrees of relationships that actually exist. (Kinsey 1923, 8)

While Gould's essay is knowing, humanitarian, and kind, it fails to notice how the categorization of organisms—including taxonomists—is rarely done simply for the sake of it. Rather, as the anthropologist Mary Douglas (1966) has described, human societies impose category systems on natural realities, creating troubling monsters, dirt, and perversion in the cracks between the category system and materiality, to warrant action and to reproduce social order. Taxonomies can be examined not only as deviations from unclear disciplinary norms. Rather, they can be read also for the social orders that they read onto the materiality of life and the resulting actions that they legitimate.[15] From this more cultural perspective on the ways that humans such as taxonomists categorize and order organisms, Gould's statement that social sciences about humans are more value-laden than natural sciences of insects is a precarious way to categorize scientists.

Like many discussions of the relationship between Kinsey's insect biology and his studies of human sexual behavior, Gould paid little attention to the distinct national ideologies that affected Kinsey's thinking about difference in both cases. And yet there is a kind of unspoken "banal nationalism" about the Kinsey studies, most obviously in their claims to study natural variation "in the human male" or "in the human female" when coupled with concern for sampling the population of the United States.[16] Indeed, both of Kinsey's research careers are characterized by impulses to counter the purification of American life of its queer individuals on behalf of the interests of the United States government. Consider Kinsey's taxonomies of American gall wasps first. Economic entomology was the first professionalized form of American entomology. Then and now, economic entomology was largely oriented toward pest control.[17] Agricultural entomology in the service of their annihilation was strengthened in the 1920s as the rhetoric and technology of military bombing techniques developed for World War I was translated into a crop-dusting "war on insects."[18] Kinsey's 1935 article in the *Journal of Economic Entomology* attempted to curtail this state-sponsored annihilation, insisting that gall wasps rarely endanger the oak trees upon which they live as parasites, and that their classification by economic entomologists among the "insects injurious to the oak" was an unnecessary error.

Not all insects require such human champions. As Eric Brown (2006) notes, social hymenoptera such as ants and bees have often been intriguing "others" for humans because they demonstrate that human intelligence is not necessary for the emergence of complex social order. Social hymenoptera also have been persistent metaphors for social order in Western thought.[19] Early-twentieth-century American entomology was no exception. Charlotte Sleigh has narrated with exquisite skill how the early entomology of William Morton Wheeler opened up new noneconomic rationales for insect study, through his new science of ant sociality: "myrmecology."[20] In the 1910s, Wheeler developed a complex evolutionary narrative about the emergence of ant colony "superorganisms" from more asocial hymenoptera such as termites and the solitary wasps that his student Alfred Kinsey would later study. In a queer narrative that decoupled the heterosexual breeding pair from evolutionary change, Wheeler attributed an ever-increasing range of insect social functions to *trophylaxis*, or social feeding. His most transparent analogy between ant and human sociality was presented in a satirical piece, written from the perspective of African Termite Wee-Wee, "43rd Neotenic King, of the 8429th dynasty of the Bellicose Termites" (1920, 124). Speaking as Wheeler's mouthpiece, Wee-Wee explained how trophylaxis had characterized evolution from primitive termite origins to stable colonial order. Emerging alternately as a critic of Bolshevism, individualism, and class societies, an advocate of eating anarchists to put their proteins to good use, an enthusiast for euthanasia, a critic and advocate of birth control, and an occasional Freudian, Wee-Wee castigated humans for the inefficiency of their non-eugenic sociality. Kinsey loved the piece and recommended it to several generations of IU students.[21]

Within the theoretical space that Wheeler's myrmecology opened up, asocial hymenoptera such as American gall wasps had import as "missing links" in the emergence of complex ant sociality.[22] Drawing on his precocious childhood scouting skills, "get-a-million Kinsey" toured the American West on long solitary field trips collecting the wasps and the galls that they made on the surface of oak trees.[23] Prefiguring Mary Douglas's paradigmatic explanation that dirt is simply "matter out of place"—such as food in the bedroom—the gall wasps that he studied exemplified a most asocial form of eating, as the gall was "a dining room and a bed chamber for the creature during its larval life," through which the adult winged wasp chewed its solitary escape to the outer surface.[24]

In his 1935 *Journal of Economic Entomology* article, Kinsey described American gall wasps as being of greater interest to taxonomists than to economic etymologists. However, as Sleigh notes (2007, 131), solitary wasps were also readily understood by amateur etymologists as "individuals" who could serve as moral analogies for individual self-improving Midwesterners. Sleigh further describes scouting as "a Spencerian form of frontier ideology" (111), and the entomological works that Kinsey wrote to explain what he had learned in the field contain within them a kind of frontier narrative for queer individuals. Those works make no reference to Frederick Jackson Turner's influential frontier thesis, and during the decades when Kinsey was pursuing gall wasps, Turner's thesis—that "what has been distinctive and valuable in America's contribution to the history of the human spirit has been due to this nation's peculiar experience in extending its type of frontier into new regions" (Turner [1920] 2011, 9)—fits Kinsey's Western gall wasps well. In particular, although Turner's description of how American national character—distinctly masculine—emerges organically when American men are separated from older European influences was falling out of favor among historians.[25] However, consider Turner's claims about how Americans are made:

> The separation of the Western man from the seaboard, and his environment, made him in a large degree free from European precedents and forces. He looked at things independently and with small regard or appreciation for the best Old World experience. He had no ideal of a philosophical, eclectic nation, that should advance civilization by "intercourse with foreigners and familiarity with their point of view, and readiness to adopt whatever is best and most suitable in their ideas, manners, and customs." His was rather the ideal of conserving and developing what was original and valuable in this new country. ([1920] 2011, 136)

An analogous narrative emerges from a paper by Kinsey in 1937 in the *Proceedings of the National Academy of Sciences*. Therein, Kinsey defined the North American gall wasps as belonging to either "continental" species in the oak-rich East or "insular" species in the Western states, Mexico, and Central America. The Eastern species were found in Europe also, were not distinctly North American, and were characterized by "intergrades"—transitional stages between species. In contrast, in the Western ecology, which Kinsey analogized to Darwin's Galapagos Islands,[26] gall wasps were confined to "insular" mountaintops where oak trees grew, and intergrades were comparatively rare. Three-quarters of the species that Kinsey claimed to have discovered through the previous two decades of fieldwork—the discoveries for which Gould laid the criticism of splitting—had emerged from this "insular" ecology. These ecological differences afforded different evolutionary narratives. Accumulative variation occurred through reproduction in the East, but the absence of intergrades in the West suggested that spontaneous mutation drove speciation in that region (1937, 8). The latter kind of evolutionary development was more likely to occur as wasps spread out to new territory:

> There is some reason for believing that most of these insular species have originated on the periphery instead of in the heart of the parental range, or, more often, have moved out of the parental range into territory previously uninhabited by that specific stock. . . . In almost every instance in insular areas the younger species occupies territory (either geographic or host) removed and even remote from the areas previously occupied by the older stocks. (1937, 10)

Kinsey concluded that this work supplanted traditional models of evolution; hybridization had been overemphasized and mutation sidelined by the artifact of human history that the European and Eastern species had been studied first. Thus the question of whether or not Kinsey had discovered many new species in the West—the matter at the heart of Gould's critique of Kinsey as a splitter—could not be simply resolved by appeal to the definition of species boundaries. Rather it depended upon the question of whether heterosexual reproduction was itself *essential* to narratives of the emergence of new forms of American life on the Western frontier.

Kinsey connected images of the American nation to his understanding of differences between people at several points in his career. His large taxonomic study, "The Gall Wasp Genus Cynips: A Study in the Origin of Species" (1929), would seem like an odd place to find such an analogy, but it is there. This large systematic study was published five years after the Immigration Act of 1924 (the Johnson-Reed Act), which limited immigration to the United States to 2 percent of the population from each national group recorded in the 1890 census. In an atypical passage, Kinsey shifted from taxonomic description to moral lessons drawn from insects for "the immigrant problem":

> It is this problem, the attainment of homogeneity out of a hybrid population, that has become the immigrant problem of the American people in the last half-century. Whether homogeneity is a biologic virtue or not, the statesmen have shown themselves good taxonomists in their insistence that we cannot become a true species until barriers are erected to protect us from continued contributions of parental stock. Whether in people or insects, the melting pot cannot blend diverse materials that pour in too rapidly. (1929, 57)

Later, in his 1939 address to the promising youth of the Phi Beta Kappa society, Kinsey gestured that the number of living organism on earth is so vast that it "would seem Utopian even to a New Dealer though it startles a mere biologist" (1971, 4).

While Kinsey's taxonomy may not have ontologized gall wasps as necessarily "naturally queer"—to use sociologist Myra Hird's (2004) memorable phrase, he used related analogies to associate the Western frontier with the natural state of male bisexuality in *SBHM*. With the exception of the chapter on animal contacts itself, animals are mentioned most often in *SBHM* in the chapter on "homosexual outlet." Animals figure Kinsey's decoupling of "masculinity" and "femininity" from the definition of homosexuality at the chapter's opening and closing. The chapter begins with a critique of the assumption that particular sex positions among animals can be labeled "masculine" and "feminine" (*SBHM*, 613–14) and ends by contesting entomological research that conflated homosexuality and intersexuality (661). Kinsey also uses the analogy of species boundaries to make homosexuality a matter of universalized importance to all rather than something that defines a distinct minority of individuals. In one of the most quoted lines from his work, Michel Foucault later analogized the historical emergence of European homosexual identities to speciation: "The sodomite had appeared as a temporary aberration; the homosexual was now a species" (1978, 43). Surely, one of the most effective oppositions to the minoritizing impulse that Foucault detected in sexology was Kinsey's description of all people as existing on a universal continuum from exclusive homosexuality to exclusive heterosexuality. In introducing this continuum, Kinsey, like Foucault drew upon the metaphor of species boundaries, explaining to his readers how his taxonomy led to the conclusion that "the world is not to be divided into sheep and goats" (*SBHM*, 639).

In Sedgwick's terms, Kinsey's was a "universalizing" epistemology of homosexuality that firmly rejected any ontological relationship between homosexuality and gender inversion. For men to enjoy homosexual sex with each other, neither a signature effeminate personality nor a feminizing hormonal pattern nor an eschewing of heterosexual pleasure was necessary. Kinsey insisted that the sexual sciences had wrongly split men into separate species because of the erroneous confusion of a few observed urban cases with the full range of natural variation. Fieldwork beyond the cities provided Kinsey with a new "discovery" that human sexual orientation was a *continuous* trait, and that male bisexuality was the ordinary balance point between exclusive homosexuality and exclusive heterosexuality, and not a category violation, a dirty impurity in the data, or a form of a more genuine heterosexuality or homosexuality. Scouting fieldwork was the taxonomic method needed to see how blurred the species boundaries were between sheep and goats; an image of manly bisexual frontiersmen provided Kinsey with a salient intergrade:

> It has also been pointed out that in certain of the most remote rural areas there is considerable homosexual activity among lumbermen, cattlemen, prospectors, miners, hunters and others engaged in out-of-door occupations. The

homosexual activity rarely conflicts with their heterosexual relations, and is quite without the argot, physical manifestations, and other affectations so often found in urban groups. There is a minimum of personal disturbance or social conflict over such activity. It is the type of homosexual experience which the explorer and pioneer may have had in their histories. (*SBHM*, 630–31)

In sum, something more than the abstract logics of taxonomy made Kinsey's two projects similar to each other. Jointly, his two projects offered an idealization of variation in American life that was not centered on heterosexuality, but that invested in banal references to frontier life. For Kinsey, queer individuals—whether wasps or WASPS—continued to evolve a distinctiveness from their European forebears through lives of self-creation in harsh ecologies, where they ate where they slept and pushed ever Westward.

The Queerness of Genius

Sedgwick described modern thought about homosexuality as littered with "ignorance effects," forms of deliberate blindness and unwillingness to know. Hitherto, my argument about Terman has been indebted to this insight—shared by Foucault and Sedgwick—that silence can be *part of* the discourse of sexuality and not just the limit or boundary of that discourse. I want to argue that Terman's surprisingly limited response to Kinsey's claims about homosexuality were—akin to his silences about masturbation and husbands' sexual performance—silences that occluded possible epistemological relationships between intelligence and sexuality. They were silences upon which the normalization of the gifted depended. Terman's review said nothing about the remarkable frequency of homosexual conduct reported by Kinsey. He only critiqued Kinsey for not making enough of a characterological distinction, by implying that "the sex chosen as partner in a sexual activity is as unimportant as one's preferences regarding food or the cut of one's clothes" (1948, 457).

Kinsey thought sex between men to be natural and widespread. When homosexuality became psychiatric grounds for exclusion from the armed forces for the first time during World War II, tests of homosexuality often assumed—as Robert Ross had done—that a distinct "homosexual personality" might allow psychiatrists to distinguish true cases of homosexuality among the troops.[27] Kinsey was critical of these army tests in *SBHM*; the prevalence of bisexual histories in his own data demonstrated the army tests' failure to exclude vast numbers of men with homosexual experience (*SBHM*, 662). Terman's thinking was closer to the logic contained within those tests. He thought that homosexuality characterized only a minority of people, who had a distinct character in some cases, and whose homosexuality could often be normalized in other cases. Indeed, those army tests often drew on the normalizing logic of measuring gender—introduced in Terman's gifted study, and expanded as a method of detecting and normalizing people suspected of homosexuality in his volume *Sex and Personality*.

Terman's response to Kinsey emerged from concerns for gifted children. A propos of Kinsey's 1952 lecture at Stanford, Terman described to Yerkes how Kinsey had joked in his lecture about the armed forces' failures to screen out homosexuals. Terman expressed to Yerkes his own pity for the gifted children whose lives he believed to have been ruined by homosexuality, and he registered his concern that Kinsey's normalization of homosexuality might be promoting the practice.[28] After *Sex and Personality* was published, Terman received occasional letters from young men troubled by homosexual feelings, men to whom he sent reassurance that their feelings were likely just passing phases.[29] Terman wrote to Yerkes that homosexuality had occurred in "some 15 or 18 in my gifted group, nearly all of whom have been wrecked in one way or another and three of whom have been blackmailed."[30]

This letter was not an anomaly. As Andrew Elfenbein (1998) has argued, Terman's thoughts were "haunted" by the possible return of the figure of the queer genius. Elfenbein describes how the notion of creative artistic genius was actively cultivated in Britain of the late eighteenth and early nineteenth centuries by authors without education, wealth, or patronage to legitimate their individual rights to earn a living from writing. The romantic genius had an eccentric persona, and was perceived to transgress norms of civil humanism that prescribed moderation and self-regulation. While private genius could be evidenced by public androgyny, the romantic construction of genius was paradigmatically male, and Elfenbein argued that eccentricity and androgyny came to signify both the possibility of a problematic private sodomy and the possibility of rarified access to nature's sublime limits. As I detailed in chapter 1, debates about the sanity of genius and the speciation of the modern homosexual of the late nineteenth century were mutually constitutive. By extending Elfenbein's discussion of how Terman was "haunted" by romantic discourses about genius, we might explain Terman's silence about Kinsey's findings of widespread normal enjoyment of homosexual sex.

Is the Gifted Child a Little Queer?

In the first chapter of this book, I described how my own psychology experiments led me to Terman and Miles's *Sex and Personality*, the seemingly "foundation" attempt to measure gendered personalities in psychology. However, *Sex and Personality* had a prehistory in Terman's study of the gifted cohort (Terman and Miles 1936, 13). Recall that Terman's earliest work drew him into conversation with the literature on genius, and that G. Stanley Hall's writings often linked the overactive mind with the risks of masturbation, inversion, and degeneracy. Terman's early move away from Hall led him to ontologize an associationist mind in which quickness led to precocious cleverness and not to precocious sin. In 1922, Terman began to instantiate an ontological relationship between giftedness and normal gender with tests of personal preference and personality. *GSG I* reported a test in which games and activities were given a "masculinity index" based on the proportion of boys and the proportion of girls—in both the gifted and control groups— who liked each one. Each *child* was then given a "masculinity" score based on the average masculinity of her or his preferred games and activities. As with the measurement of "intelligence" as IQ, Terman had developed a neo-Galtonian way to measure individual differences in "masculinity" in which there was no meaningful zero point to the measure, and everyone had each property in an amount that existed only by comparison to others.[31] As feminist and psychoanalytic readers might have predicted, Terman's masculinity was signaled by presence and femininity as absence in the original test; the masculinity of activities ranged from $+24$ ("playing with tools") to $+2$ ("playing with dolls"). Comparison of the gifted and control groups showed that gifted boys, but not girls, were very slightly more masculine than the controls (Terman 1925, 437). Terman expressed relief that "these correlations offer no support to the popular belief that the gifted boy is effeminate in his play interests" (404).

The ghost of queer genius was not exorcised as easily as that. The gifted project was not simply a neo-Galtonian study in which participants' individuality dissolved into aggregate scores. In addition, gifted children were also individualized, in histograms that made individual scores visible, and in verbal case histories. In *GSG I*, a histogram of the "masculinity scores of the gifted and control boys" shows one nine-year-old gifted child, categorized as a boy, but clearly much less "masculine" than any of the other gifted or control children so categorized (411; see fig. 2). No commentary on the outlier in this graph accompanied it in the text.

Such children as the one represented as an outlier in figure 2 were Terman's prompt to begin a larger empirical project aimed at the normalization of gender via its neo-Galtonian measurement. Terman secured grants from the CRPS in 1925 and 1926 for $1,500 and $2,900 respectively for the "Development of Tests of Masculinity-Femininity Traits in the Non-Intellectual Aspects of Mentality." He was granted a further $21,000 between 1927 and 1933 for the study of "Sex Differences in Non-Intellectual Mental Traits."[32] In 1925 and 1926, Terman and his students began to develop the masculinity measure using word associations, ink-blot projective tests, general knowledge tests, and questionnaires about emotions and "introvertive responses" (Terman and Miles 1936, 18–51). In March 1927, Terman invited his former student Catharine Cox back to Stanford to work on the project. Cox had published her Ph.D. in 1926 as *Genetic Studies of Genius II*, extending Terman's study of Galton's childhood genius to three hundred historical figures.[33] As a reminder of the limited opportunities for smart women in academic psychology, Terman wrote to Cox that he had recommended her for a faculty position at Minnesota, "but I suppose there is the old question whether a woman will be given a fair chance at it."[34]

Research on the measurement of "masculinity-femininity" and the follow-up studies of the gifted proceeded simultaneously at Stanford in the late 1920s and often intersected in psychologists' careers. For example, the second book on the gifted cohort, *GSG III*, was largely written by the

TABLE 153a.
MASCULINITY RATINGS OF CONTROL AND GIFTED BOYS, BY AGE

Rating	6-7 G	8 C	8 G	9 C	9 G	10 C	10 G	11 C	11 G	12 C	12 G	13 C	13 G	14 C	15-17 C	Total 8-13 C	Total 8-13 G
20							1			1						1	1
19					1		2			0		1				1	3
18			3		0		1	2		0	2	2				4	6
17	1		2		5	3	7	3	12	2	4	6	3			14	33
16	2		10	4	7	2	21	5	16	7	15	16	7			34	76
15	2	2	8	3	15	7	23	7	25	7	15	14	6			40	92
14	2	3	15	4	14	9	12	8	15	8	4	2	2			34	62
13	7	3	3	8	3	8	1	7		4	1	1				31	8
12	2	1			3		2		1							1	6
11	1		1				1										2
10						1											
9																	
8																	
7																	
6																	
5					1												1
Total	17	9	42	19	49	30	71	32	69	29	41	42	18	34	30	161	290
Mean	13.71	13.67	14.95	14.16	14.63	14.27	15.38	14.84	15.32	15.00	15.56	15.81	15.61	15.65	16.07	14.90	15.22
S.D.	1.57	0.94	1.45	1.18	1.90	1.46	1.48	1.49	1.08	1.49	1.04	1.12	0.89	1.24	1.09	1.49	1.45

Fig. 2. The Masculinity of Gifted and Control Boys, including an Outlier (*GSG I*).

precociously gifted psychologist Barbara Burks (1902–43). Burks pioneered the statistical techniques which continue to ground the trenchant nature/ nurture debates about intelligence in American psychology.[35] *GSG III* reported on an empirical study of seventy-five boys and seventy-two girls of the gifted group who were compared with boys in girls from a San Jose high school. The gifted boys were only slightly more feminine than the controls, but the greater masculinity of the gifted girls was "established with practical certainty" (Terman, Burks, and Jensen 1930, 156). In the interpretation of these differences, anxieties about queer genius were focused far more on atypical femininity than masculinity:

> There is no evidence that invert tendencies are significantly more common among men of genius than in the general male population, but there is reason to believe that the incidence may be higher for the general female population. The evidence suggests that there is at least a little basis for the exaggerated opinion quoted from Goncourt by Lombroso that "There are no women of genius; the women of genius are all men." (Terman, Burks, and Jensen 1930, 161)[36]

GSG III was not a feminist text that argued for the flourishing of queer children who had been sexed-as-female. Rather, androcentric interest in gifted children who were sexed-as-male is evident in the case histories of two gifted children. With an IQ estimated at 160, "Roberta," "our most masculine girl," was "exceptionally dexterous with crayons and play tools" and later constructed toy airplanes and other mechanical devices. Roberta was interested in war games and movies, and "the mother finds this strange military interest distressing." However, the IQ psychologists directed attention away from the output of Roberta's creativity: "even more interesting than the actual nature of her mechanical output are the many angles of her personality which have been reflected in tests and personal reports." Instead, Roberta's character stimulates nature/nurture questions that yield no easy answers:

> What one would like to know about Roberta is whether her masculine and scientific bent is the result of an early fixation on the father, perhaps accentuated by his death when she was still a young child, or whether it is but the manifestation of a physiological endowment that predisposes her to masculine pursuits.[37]

In contrast to Roberta, "Renwick"—a gifted musician with an IQ of 150— had "the most extreme case of feminine personality bias among our gifted boys" (329) in addition to notable musical talents.[38] This description leads me to believe that "Renwick" is a pseudonym for the child represented by the outlying "masculinity score" in the histogram in *GSG I*. At age eleven, Renwick's parents installed a pipe organ in their home. Renwick was performing concerts at age thirteen and had, at age fifteen, been appointed a church organist and choirmaster, and was studying counterpoint and composition actively. Terman cites "the forecast of professional critics" that Renwick will have a "brilliant future in music" (329).

Unlike Havelock Ellis, Terman did not consider musical composition a particularly masculine form of intellectuality. Rather, "male musicians and artists average decidedly less masculine on the Masculinity-Femininity Test than unselected males" (328). However, with regard to Renwick, unusual genius was less of a worry than unusual gender:

> Paralleling Renwick's early musical interest was an intense interest in playing "dress up." Here he inclined toward the feminine. He composed playlets and operas, always casting himself for the leading feminine rôle. He played with dolls and amused himself by dyeing clothes and designing feminine garments. He refused to play with boys, notwithstanding punishment at home and merciless criticism at school. (328)

The partial accounts of Roberta, Renwick, and other gifted children with "inverted" tendencies make these children's lives difficult to locate neatly within such twenty-first-century taxonomies such as "lesbian, gay, bisexual and transgendered." They stimulated Terman to initiate the study of "gender" as a personality attribute in a more general sense, culminating in the book *Sex and Personality* (1936), which Kinsey would critique.

From Minoritizing to Universalizing Masculinity-Femininity

The careers of other psychologists whom I have already mentioned were also involved in the writing of *Sex and Personality*. E. Lowell Kelly gathered the data that would inform the conclusions about male homosexuality while he was struggling to understand his own marital difficulties. The manner in which Kelly was given his duties speaks to the ways that gender, sexuality, and intelligence were "in the making" in this project.[39] Terman brought Quinn McNemar and E. Lowell Kelly to his office and confronted them with a choice; one would go to San Francisco to find homosexual subjects, and the other would remain in Stanford to conduct the statistical analyses. Before Kelly could speak, NcNemar put forward his own statistical services, and Kelly was consequently given the fieldwork.[40] In the division of scientific labor in this project, masculinity was performed by adhering to abstract analytic work and by avoiding contact with homosexuals. The male psychologist with the quickest mind was rewarded with the intellectual job and the slower one was condemned to associate with homosexuals. In the resulting volume, the most masculine occupational groups were men who engaged in abstract analytic work—the engineers and architects. The "passive male homosexuals" that Kelly studied in San Quentin prison were described as the most feminine group.[41]

Both of Terman's biographers emphasize that *Sex and Personality* was a new departure from his earlier work on intelligence.[42] This theory of gender bore a strong impression of the neo-Galtonian epistemology of Intelligence testing. The text of *Sex and Personality* compares the logic of normalizing the M-F measure to the development of an IQ test (64) and compares its utility in detecting children with inverted tendencies to the utility of IQ tests in detecting the gifted (468). While the Stanford-Binet test was to be used to help the precocious intelligence of gifted children to flourish yet further, the M-F scale was useful apropos of boys who develop differently, because "preventative measures might be found that would direct their sexual development into normal channels" (264).[43] Masculinity-femininity scores—where masculinity is always denoted by positive numbers—were also positively correlated with IQ scores and educational achievement (139), the scores of college women were decidedly more masculine than their less educated peers (146), and education was analogized as "mental masculinization" (139). Fourteen case histories of "delinquent girls" showed that those who were "masculine" had above-average intelligence, while those who were "feminine" did not (321–41). Indeed, another team of personality psychologists, working at Stanford at the same time, produced a very different account of the "masculinity-femininity" dimension. In their later work, those psychologists were explicitly critical of Terman's conflation of masculinity with white-collar masculinity.[44]

A third account of the child known as Renwick appeared in *Sex and Personality*. A simplified version of the histogram from *GSG I* was reprinted in the introductory chapter to this volume (see fig. 2).[45] We are now told that an assistant had found the masculinity scores of one of the gifted boys, identified only as "X," to be so feminine that it suggested that X was a girl who had been classified incorrectly: "The score was accordingly checked for error . . . but no error was found" (14). X was figured as a queer gifted child indeed; "one of his favorite amusements was to dress himself as a stylish young woman, apply cosmetics liberally, and walk down the street to see how many men he could lure into flirtation." As X "naturally provoked speculation in regard to the course of development that would ensue," the Stanford researchers sent occasional letters to X's mother asking her for details. Although these letters were "couched in general terms to avoid the risk of causing offense or shock," the researchers appear to have underestimated X's mother's intelligence of them. She used one letter to confront X, and he revealed his sexual interests in other boys to her as a consequence (14–15). However, the psychologists found cause to hope for X's normalization. They speculated about "excessive attachment to his overcherishing mother," consistent with Terman's conclusion in his later book on marriage about the Oedipal origins of homosexuality. Indeed, like Freud's da Vinci, Terman thought that X could transcend the bounds of ordinary human sexuality through genius: "X is an example of the highest type of mental sexual inversion; he has high principles, is passionately devoted to his work, and seems to have rejected all overt expression of his homosexual inclinations" (15). The promise of intellectually precocious youth might yet be sublimated; the poison of homosexuality might become the gift of genius.

In chapters 2 and 3, I examined Terman's unwillingness to link creative intelligence to masturbation and to engagement in mutually satisfying heterosexual sex respectively. *Sex and Personality* contains many such failures to recognize the intelligence of people categorized as both gender-inverted and as male. Terman and Miles utterly failed to consider how X's feminine behavior, avoidance of stigma through drama, years of surviving the scrutiny of psychologists, or successful flirtation with sailors are cases in point.[46] The book's later case histories of "passive male homosexuals" co-authored by Terman and Kelly contain similar misses. While not abnormally high in IQ, the "passive male homosexuals" found intelligent ways to live that were *not* marked as smart in the text; one dressed as a girl and performed for day laborers to earn dimes, and another became a paid female impersonator (285–91). The M-F test was also re-normed to develop a measure of "inversion" that distinguished the responses of this group from those of ninety-eight male junior high school students. Here also, the explanations of differences are telling. Better knowledge of composers among the "passive male homosexuals" than among the students was attributed to "the fact that so many of the inverts had musical interests" rather than superior knowledge. Better detection of grammatical errors among the passive male homosexuals was devalued by its attribution to their being "more fastidious in their language as in other things" (270–71). Unsurprisingly the "passive male homosexuals" were not trusted as sources of scientific intelligence either, particularly with regard to their belief that homosexuality is inborn. Terman and Miles insisted instead that the evidence of the case histories suggested "impressive evidence with respect to the environmental causes of sexual inversion" (284).

However, attributing these views to "Terman and Miles" may be problematic. Relations between Terman and Miles became strained over the writing of *Sex and Personality*. The two fell out, and at one point Terman wrote to Yerkes that "as a result of the numerous disagreements, and perhaps also misunderstandings, our conferences during the last year Mrs. Miles was here, were mostly futile and always left her, and sometimes myself also, very upset emotionally."[47] However, Miles's science succeeded where Terman's had failed. He struggled to construct a sample of lesbians to which the M-F test could be applied, but a comparison of "18 women inverts" with 988 women with high-school and college educations appears at the very end of the book, and the account of these women was written by Miles. While the neo-Galtonian logic of the comparisons is consistent throughout the book, Miles did acknowledge the contribution of lesbian researcher Jan Gay in "securing the cooperation of these 18 women."[48] In other words, this book was skewed by a distinct fear about gender inversion among people designated the male gender.[49]

"Everyone Considered Henry as Queer"

X/Renwick was not the only child that Terman ontologized as gifted who would later evidence a little queerness. Nor were the "passive male homosexuals" the only prisoners in San Quentin who would preoccupy Terman. In his book on Terman's gifted children, Joel Shurkin (1992, 296) concludes that the composer Henry Cowell (1897–1965) was the exceptional case among Terman's subjects because he achieved true greatness. Cowell was not part of the gifted cohort per se but had been identified as a gifted child by Terman some years earlier. In 1910 Cowell, was working as a janitor in the public school that he had earlier left because of bullying and had just ceased making his living by begging and selling ferns to Stanford faculty members, when he first met Lewis Terman. At the time Cowell was being raised by, and was supporting, his mother, Clarissa Dixon. Dixon was a modern divorced woman who distrusted formal schooling, wrote a novel centered on love between girls, and who instilled a bohemian spirit in her son. Later, Terman reported Cowell's Stanford-Binet score to be 132, just below the usual cut-off used for inclusion in the gifted category. Terman considered this score an underestimate but noted that at this point "everyone considered Henry as queer, not to say freakish" (1921, 248–49).

Michael Hicks's biography of Cowell (2005) narrates how his background in Californian bohemian subcultures had lasting determinative effects on his creative output. In 1912, Clarissa Dixon met Ellen Veblen, one year after her divorce from Thorstein. On meeting Veblen, Dixon wrote that "I knew I was not too old for another love," but Hicks's biography says little more about the relationship between these two modern women. Ellen Veblen financially supported Clarissa Dixon and Henry Cowell, moved them to her cottage in Carmel, and ultimately made Henry the executor and benefactor of her will. In 1914, Stanford professor Sam Stewart organized a recital in Palo Alto in which Cowell played his original compositions, which led to local newspaper descriptions of a "youthful wonder" with the "charm of genius," and the launch of an educational fund for Cowell.[50] The fund supported two years of education at the University of California, Berkeley. After his mother's death in 1916, Cowell traveled to New York, where he met the composer Leo Ornstein. Each was impressed with the other; Cowell told Terman that he lay awake at night contemplating Ornstein's genius,[51] and Ornstein described Cowell's juvenile works as "the most interesting compositions I have seen by any living American."[52]

In his account of the young Cowell in *The Intelligence of School Children*, Terman displayed his typical preference for scientific professions when he lamented that "if he [Cowell] obtains fame as a musician, his biographer is almost certain to describe his musical genius as natural and inevitable, and to ignore the scientist that he might have become" (1921, 251). Cowell exemplifies the apparent contradiction in Terman's relationship with the gifted children he sponsored. On the one hand, in *The Intelligence of School Children*, Cowell's life became evidence of how natural genius would emerge in spite of educational disadvantage. On the other, Terman supported Cowell's education deliberately lest his genius go unactualized.[53] The contradiction is about temporality. The achievements of gifted people seem inevitable in retrospect, but future accomplishments seem dangerously precarious and to need particular support.

In the 1920s, Cowell appeared to be making good on the promise of youth, albeit in music rather than in the sciences. He achieved fame on both sides of the Atlantic as an innovative composer of distinctly modern music. Cowell was patronized by composer Charles Ives, and was far removed from the circle of queer composers who wrote in a simpler harmonic idiom that became associated with a distinctly American musical tradition.[54] Critics often compared Cowell and the composer Leo Ornstein in the early 1920s because both explored modern forms of harmony which involved playing several contiguous notes on the piano simultaneously; Cowell would come to call such sound tones "clusters." After meeting Ornstein in New York and after his mother's death, Cowell joined the Temple of the People in Halcyon, California, a community of theosophists led by the Irish poet John Varian. There he inhabited a house that he was bequeathed by Ellen Veblen. At Halcyon, Cowell composed experimental works to accompany Varian's mythological dramas. In 1922, shortly after the death of his fiancée Edna Smith, Cowell began a relationship with a twenty-one-year old man at Halcyon which came to the attention of the police, and he managed to avoid arrest only through referral to the counselors at the Temple.

Terman continued to invest in Cowell's persona as a genius as his fame grew, and Cowell became both an author and a subject of the psychology of genius. In 1926, Terman introduced Cowell's paper "The Process of Musical Creation" in the *American Journal of Psychology*. Citing another critic, Terman wrote:

> Concordances of many close-lying notes have been used by Leo Ornstein since he wrote his Dwarf Suite; and Percy Grainger calls for notes struck from the strings inside the box of the piano in one of the Nutshell movements, but it has been left for the young Californian to demonstrate completely the quality of sound to be produced on concert grands by deliberate application to the keyboard of muscles other than those in the finger-tips, and by the application of the fingers to the strings themselves.[55]

Cowell continued with an account of his own creative process. He described how, at an early age, he desired to hear music frequently, and how his "sound mind" then mentally recreated music. At times "glorious sounds leap unexpectedly into my mind—original melodies and complete harmonies such as I could not conjure forth at will." At this early stage "I had at first not the slightest control over what was being played in my mind at these times; I could not bring the music about at will, nor could I capture the material sufficiently to write it down" (Cowell 1926, 235).[56] Later, Cowell learned to control, and to write, the music in his sound mind. But the results were disappointing: "No player can play as perfectly as the composer's mind . . . no other instrument is so rich and beautiful, and . . . only about ten percent of the musical idea can be realized even at the best performance" (236).[57]

Hicks (1993) describes this paper as part of a larger strategic attempt by Cowell to craft a persona for himself as a romantic child genius, a persona that Hicks gives his readers some reason to question. As the "clusters" use all of the notes within a single octave, the tones in between are difficult to hear. Thus, far from advancing the frontiers of harmony, clusters can be experienced simply as percussive noises. Furthermore, Cowell hardly "discovered" the cluster method, as many small children who sit at a piano will play clusters spontaneously as they explore the piano with their bodies. As such Hicks argues that their "discovery" by Cowell "was something akin to Columbus's 'discovery' of a land already populated" (1993, 432). Cowell's claim to originality and genius rested not on new playing practices but on the redefinition of existing practices as a modern intellectual method. While he achieved considerable fame for these innovations, they were viewed with cynicism by some of his contemporaries.[58]

Cowell's *American Journal of Psychology* article described the cluster compositions as being composed "in mind" years before they were written down. Describing this ideal composer's mind, the article psychologized Cowell's account of his own childhood creativity. Cowell's introspections were more consistent with Ellis's description of the male composer's mind as one that is overcome with energy than with Terman's descriptions of IQ as akin to electricity.

> The experience of being in the throes of musical creation is distinctly an emotional one; there is a mere semblance of the intellectual in being able to steer and govern the meteors of sound that leap from the mind like volcanic fire, in a glory and fullness unimaginable except by those who have heard them. (1926, 236)

Indeed, against Terman, Cowell negated the idea that composition is an intellectual process that IQ might measure. Rather, it is the emotional experience of musical ideation which Cowell foregrounded here. By investing in Cowell's persona, and presenting his introspective report of composition to his psychology colleagues, Terman bound the fate of his larger project to normalize gifted children to the life and work of this "freakish" and audibly avant garde figure. Cowell was offered the position that romantic discourses and Galtonian normalization both affords—that of talking about one's own genius as a form of idealized intelligence in autobiographical terms.

Marcel Mauss insisted that gifts bind giver and receiver together because gifts remain active after they are given. Cowell's account of himself in 1926 made good on the "promise of youth" that Terman hoped that IQ tests would detect, but in an ironic way. Cowell spoke to psychologists in this article, as someone who had been designated a child genius by Terman's IQ test. By speaking from within a romantic idiom, Cowell authored an account of his own mind that spoke to psychological possibilities that went beyond Terman's theory of Intelligence. Rather than being a "confession" that incites normalization from the psy- disciplines, one could consider that Cowell faithfully repaid Terman's gift with this challenge to the intelligence of Intelligence testing. Cowell's gift invited Terman to consider a wider range of ontologies of smart minds than his tests of "Intelligence" could discern.

As the interconnected lives of Terman and Cowell went on, homosexuality came to define the difference between making good on the promise of youthful giftedness and reneging on it. In April 1936, a month before *Sex and Personality* was published, Cowell was arrested for having oral sex with one of the seventeen-year-old boys who frequented the pond behind his Menlo Park home.[59] Cowell confessed freely, and Californian newspapers described him as a child molester; the trial was analogized to Oscar Wilde's,[60] and Cowell's infamy exemplified the growing panic about "sex crimes" in the state and in the nation.[61] Cowell entered San Quentin on July 8, 1936. His stepmother, Olive Cowell, began an appeal for his release, and Terman wrote a letter in support. In that letter, Terman argued that Cowell was not a "true homosexual" but merely had delayed heterosexual adjustment.[62] As with da Vinci's mother and X/Renwick's mother, the bohemian Clarissa Dixon was blamed for her son's unnatural homosexuality. Yet this hope of a future sublimation failed to affect Cowell's fate. Olive Cowell was unsuccessful in convincing the California Board of Prison Terms to review Cowell's fifteen-year sentence. Unlike the earlier interventions of the counselors at the Temple of the People, the opinion of the psychologist from Stanford did not obviate incarceration. Cowell served four years of his sentence.[63]

I have wandered across frontiers and crossed imaginary volcanic fires to answer the question with which I began this chapter. As Terman's account of Cowell, or of X/Renwick and of the "passive male homosexuals" in *Sex and Personality* demonstrate, he was ontologically invested in a theory of male homosexuality as a form of gender inversion, and in the potential malleability of that homosexuality, particularly the form that might exist among those gifted individuals capable of sublimation. In the year before Kinsey's *SBHM* was published, Terman published the fourth volume of his *GSG* series with

Melita Oden. The grown-up gifted and control children were asked questions about homosexuality, and eleven gifted men and six gifted women became "known, or believed" to have a history of homosexuality (Terman and Oden 1947, 120). Consistent with Terman's letter to Yerkes about Kinsey's talk at Stanford, in this volume it appeared that some fifteen to eighteen of the gifted had histories of homosexuality.

Although these individuals returned the possibility that the category of giftedness was haunted by the figure of the queer genius, Terman and Oden did not despair of these results. They shifted the frame of comparison for the gifted away from an ideal heteronormative standard in which homosexuality does not occur at all toward one in which the gifted were adequately adjusted because they were comparatively less likely to be queer than average. To make this shift, Terman and Oden cited two sex surveys—Katherine Davis 1929 book on women and Alfred Kinsey's 1941 paper on men—as showing that homosexual behavior occurred in about half of the sexual histories of women and men respectively. The commonness of homosexuality in those studies showed that "an overwhelming majority of persons who have had homosexual experiences make fairly easy transition to heterosexuality, but for a considerable number the transition is less easy" (Terman and Oden 1947, 121).

I began this chapter with an attempt to explain why Terman did not speak about Kinsey's remarkable finding about the commonness of homosexual experience among American men. As this use of Kinsey's 1941 paper in Terman and Oden's 1947 book demonstrates, Terman was not troubled by the evidence of widespread homosexual conduct in the population that Kinsey's work had already begun to reveal. Rather, Terman's defense of the gifted had come to depend upon that projection of homosexuality onto the national population. Terman did not keep silent about Kinsey's claims about homosexuality because he was prudish, vengeful, jealous, or uninterested. Rather, the remarkable commonness of homosexuality that shocked many readers in 1948 had already become one of several ways that Terman managed the continued haunting of the category of gifted people by queer individuals.

5

Gentlemen and Horse Traders

Critical histories of the IQ movement often imply that intelligence testers are less smart than they imagine themselves to be. Sometimes the definition of intelligence is critiqued as positively unconscious of particular kinds of intelligence, such as "social" or "emotional" intelligence. In previous chapters, I have similarly argued that Terman's writing resisted the possibility that IQ might be predictive of men's various sexual practices and desires. By so arguing, I have been calling attention to counterfactual alternatives to the conclusions that Terman drew from the data that his assistants gathered. I have also been arguing that Terman's Galtonian project—of positioning the gifted cohort as ideal kinds of people—required his stopping short of interpreting relationships between intelligence and sexuality. If my criticisms of Terman's subjectivity are taken seriously, they suggest that the definitions of "intelligence" now current in psychology are to some extent an expression of Terman's personal beliefs about sexuality and gender. Moreover, many psychologists and others would have very good reason to question continued investment in Terman's beliefs on these and other matters. If I am right about Terman, are IQ tests then less valid? Ought we limit the powers of the state to use his and others' tests to assign children to special education classes or to enact the death penalty on vulnerable perpetrators of crime? In this chapter I raise the question of the relationship between attributions about a scientist's character and the trustworthiness of scientific knowledge resulting from that scientist's efforts. How can knowledge be made to appear independent of the character of the scientist who produces it?

As I do, my attention will continue to shift away from Terman's ways of working to Kinsey's. As noted in chapter 1, James Jones's biography of Kinsey was but one of several texts which linked the dubiousness of Kinsey's science to a description of his sexual interests and character. Over the decades, historians have expressed nothing but ambivalence about the extent to which Kinsey's work was "good science." Vern Bullough judged that Kinsey's work was "of such mixed quality that it is difficult to separate the kernel from the chaff, and still more difficult to maintain any perspective during its perusal" (2004, 281). Kinsey's work has been praised as "brilliant,"[1] and his solution to sampling problems as "ingenious."[2] Kinsey has also been criticized as failing to understanding techniques such as random sampling, failing to generate rapport with working-class participants, failing to standardize his interview methods, and failing to affect the stance of a "professional stranger" in the interview context.[3] In the 1970s, historian Regina Morantz described Kinsey's work as presenting new insights about class differences, noting that "few commentators . . . have explored the implications of Kinsey's discovery that men from different classes exhibit divergent sexual behavior" (1977, 585). In the same year, Paul Robinson (1977, 46) described Kinsey as making the fundamental discovery that sexuality was influenced by class. More recent historians have categorized Kinsey as a lumper rather than a splitter, shifting emphasis from the class differences that he claimed to have found to the racial homogeneity of his samples. Miriam Reumann (2005, 4) critiqued Kinsey for emphasizing class differences between men's sexual behavior and occluding race differences by publishing only his data on White Americans. Sarah Igo (2007, 220) described the Kinsey reports as contributing to the imagination of a common national public among *White* Americans, extending the notion of the "average American" to sexual behavior. I agree with Morantz that there has been a lack of attention to class in the Kinsey literature, and those class differences are most elaborated in *SBHM*.

An Educated Intelligence

If my earlier chapters have argued that sexuality is often a topic of Terman's research on intelligence, this one will change tack by drawing out matters to do with intelligence and education from Kinsey's *Sexual Behavior in the Human Male*. On the book's first page, Kinsey described the desire of "an increasing number of persons" to "bring an educated intelligence" (3) into the consideration of sexual politics, a desire that his volume was to address. Subjects and researchers had contributed to the project, creating "a spreading knowledge of the existence of the study among tens of thousands of people, and of an increasing understanding of its significance" (10–11). In the summer of 1938, Kinsey chaired a committee of seven faculty who offered a twelve lecture non-credit "marriage education" course to senior students. Like many marriage educators of the 1930s and 1940s, Kinsey was moved by the lack of direct information about human sexuality that he might present to his students, and of the fears, shame, and misinformation that they disclosed to him in private conferences.[4] Kinsey began to gather data on his own students, and had taken sexual histories from fifty by September of that year. While the marriage course was popular among students, powerful actors within and beyond the boundaries of the campus considered both the marriage course, and the sex study, to be immoral. Indiana University president Herman Wells offered Kinsey the choice to continue to educate about sex or to continue producing knowledge about it through research. Kinsey chose the latter option and began to expand his research off campus, gathering diverse sexual histories, initially in Indiana's prisons and in Chicago's homosexual subcultures.

During this early phase of the project's development, Kinsey formed a pivotal, if brief, friendship with biostatistician Raymond Pearl, who visited IU for six weeks. Pearl's influence on Kinsey was considerable. While Kinsey was studying gall wasps in the field in the early 1920s, Pearl was placing insects in analogical relationships with human populations in his Johns Hopkins biostatistics laboratory. Along with his junior colleague Lowell Reed, Pearl put the rapid breeding cycle of fruit flies to work to generate natural laws about the biology of death and extinction within populations that could traverse boundaries of species being between human and insect populations.[5] By the late 1930s, Pearl's biostatistical approach had became marginalized within population science, and Pearl had become a public critic of eugenics and an advocate for the social determinist thesis that access to contraception—not natural laws—determined demographic variation in the United States.[6]

Pearl's influence on Kinsey's work was threefold. First, Kinsey described patterns of sexual behavior in his work as replicating those already observed by Pearl, including that enduring myth, popularized by Woody Allen, that the peak of sexual activity naturally occurs far earlier in men than in women (*SBHM*, 222). Kinsey also cited Pearl's population studies to support the normative nature of his age distribution (76), the rates of marital intercourse he observed (193), and the high rates of sexual activity that he observed among farmers (451).

Second, Pearl influenced the way that Kinsey recorded data. Together with his biology colleague Robert Kroc, Kinsey initially developed a standardized questionnaire for interviewing students about their sexual histories. Kinsey quickly abandoned this method and ultimately argued that "the most serious error" in past sex research was the overuse of questionnaires, "used because they are easier to administer and they save time" (*SBHM*, 31). Pearl advised Kinsey to record interviews in situ so as to remember them more faithfully,[7] leading Kinsey to develop his own code for recording sexual behavior. Kinsey and his interviewers used recording sheets in which "each aspect of the sex history is recorded in a particular block or portion of a particular block,"[8] adapting a practice of using special locations to index meaning from field methods used earlier in the gall wasp studies.[9] Different symbols—some of which were "standard practice in biology, chemistry, physics, and other sciences," and some of which were made up for this specific study—recorded acts, ages, and frequencies in different locations of the response sheet.[10]

Bruno Latour has described scientists as enthusiastic inscribers of natural events that occur both inside and outside of laboratories. While empirical science relies on the idea of observing nature directly rather than trusting textual authority, Latour notes that scientists ironically focus attention on *inscriptions* of nature rather than natural phenomena themselves. Scientists are busy at work "turning rats and chemicals into paper,"[11] and those same scientists often dispose of the material rats and chemicals once the inscriptions on the paper have been harvested. By relying on a system of turning speech acts into writing, Kinsey was not acting as an atypical scientist.

As this chapter will engage the question of whether readers have the capacity to imagine events which they have not witnessed themselves, let's imagine the fate of one such hypothetical inscription. A young imprisoned man tells Kinsey that he has never had a nocturnal emission. Dr. Kinsey acknowledges the refusal and asks the young man when he first experienced the sensation anyway. The young man answers that he was twelve when such a thing first happened. Kinsey makes a note by writing a symbol in a section of the General Data Sheet No. 358-230—"a standard Keuffel and Esser product" (*SBHM*, 71)—drawing down from memory his knowledge of the association between such symbols and sexual acts. But Kinsey makes no mark on the paper to record the young man's hesitancy or his own prompt. Once the interview is completed, Kinsey carries the inscription away from the prison back to Bloomington. There it becomes one of the "sexual histories" that are analyzed there, objective records of people like the young prisoner whose physical bodies are even harder to move than the tiny gall wasps who resisted experimental science.[12]

Latour (1990) calls such scientific inscriptions as Kinsey's coding sheets "immutable mobiles" that usefully allow scientists to draw together diverse information in variable ways, to carry that information away from a place, and to apparently extract the essence of the thing itself. Kinsey valued this drawing together of lives onto single data sheets for precisely such reasons: "If the record is confined to a single sheet, it is possible to correlate any item with any and every other item by a rapid sweep of one's eye over the page of simple and precisely placed symbols" (*SBHM*, 71; see figure 3). In Blooming-ton, the data sheet would join others and inform the ways that holes were punched out of cards that the IBM Hollerith machine—purchased by the Kinsey group in February 1941—would recognize.[13] Beyond "a few routine

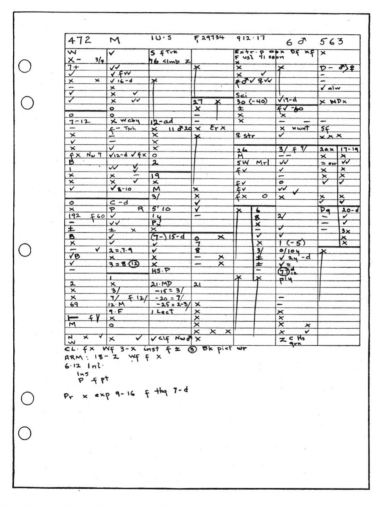

Fig. 3. Sample Sexual History in Code. *SBHM*, 72.

95

tabulations of non-sexual items" (*SBHM*, 45), all operations of the Hollerith machine were performed by a member of the Kinsey team. Clyde Martin had failed to convince Kinsey that he could be made into a convincing interviewer, and Kinsey set him to work transforming the machine's output into his own hand-drawn graphs.

Latour (1990) emphasizes that inscriptions are not the same as the things that they inscribe. As a result, the claims of empiricists to be studying nature itself rather than textual authority become precarious when scientists come to spend much longer looking at the inscriptions they develop themselves than at natural phenomena. Kinsey did not standardize the *words* of his interviews, only the range of questions that the interview must address. Many words that would have passed between Kinsey and the young prisoner were not recorded on the General Data Sheet. This point introduces the possibility of Kinsey's unchecked, untrustworthy, personal influence upon the "sexual histories" that the data sheets are taken to faithfully represent.

Indeed, the very class diversity that Kinsey's sampling aimed to discover increased the dangers of his having missed how his choice of words might have led to the telling of different sexual histories. Tellingly, Kinsey depended upon judgments about intelligence in choosing how to proceed in interviews. Rapport was established differently for different people because "poorly educated and mentally dull individuals" needed to be told that "doctors need to know more about these things. They need your help so that they can help other people" (*SBHM*, 36). On the other hand academic men "saw broad, basic principles emerging from the study" (36). The interviewer was required to make judgments about each person's "social level" in situ, and to select vocabulary accordingly, using "the clinical terms" for college graduates and slang terms for "lower level individuals" (52). One of Kinsey's techniques inspired particular doubt. "Proving the answer," was Kinsey's method of asking questions that presumed that the participant had enacted the sexual behavior in question, as with the hypothetical prisoner questioned about his nocturnal emissions above.[14] When an interviewee claimed that she had not engaged in a particular activity, Kinsey advised that the correct response was to ask a follow-up question that assumed the act had taken place. "'Yes, I know you have never done that, but how old were you the *first* time that you did it?" (55, emphasis in original). Again, assumptions about intelligence informed the use of this technique; "proving the answer" was most useful "with uneducated persons, and particularly with feeble-minded individuals" (55).

The status of these inscriptions as *representations* became highlighted in Kinsey's debate with Terman. Consistent with Morantz's and Robinson's points about the emphasis on class differences in Kinsey's work, Terman focused on claims about class in the third of his four critiques, and he raised much more general skepticism about the trustworthiness of Kinsey's methods in the process. Kinsey had quantified both the "educational level" and the "occupational class" of his participants, and the men he described in *SBHM* more often reported upward than downward class mobility (*SBHM*, 417–18). He concluded that sexual behavior not only varied by "level" but also that differences in sexual histories emerged early in men's lives and could be used to predict patterns of later class mobility: "the sexual history of the individual accords with the pattern of the social group into which he ultimately moves, rather than with the pattern of the social group to which the parent belongs." Consequently, "an individual who is ever going to depart from the parental pattern is likely to have done so by the time he has become adolescent" (419).

Terman could not imagine how these patterns might emerge faithfully from the sexual histories without the undue influence of the interviewers. Why, he asked (1948, 427) would those fifteen-year-old boys in Kinsey's "social class 4," whose fathers were skilled workmen, be masturbating more and having less intercourse if they were destined for college education rather than a life as a "semi-skilled workman" *after* these differences in sexual behavior had emerged? Once again, the matter between these two centered on the question of what sexual precocity signified. Again Terman was resistant to associate sexual precocity with life success. He pointed to "the possibility that subjects in the various social classes may not report with equal accuracy their sexual activities during the early teens" (454)—introducing a doubt about the extraction of life histories into data sheets with the interview technique. He was not alone in his skepticism. Psychologist Albert Ellis (1948) spotted a further irony; Kinsey had emphasized class differences in the social norms for sexual conduct. Lower-class people rationalized their sexual behavior on the basis of what is considered natural, but upper-class people saw morality in sexual terms (384–85). However, Kinsey failed to consider that such different norms could influence how people from different classes would talk about their sex lives to researchers.

Such criticisms about the specious nature of these class differences were potentially damning. If the class differences were a consequence of the interviewers' treatment of different kinds of men, then the zeal and perseverance Kinsey had devoted to surveying vast numbers of different people was a wasted effort. Terman's criticism threatened to disassemble the large aggregate of data that Kinsey had drawn together. Robert Kohler (2002) locates Kinsey as a fieldworker in the context of the epistemological space between field- and lab-based biology in the early twentieth century. While labwork is uniquely committed to placelessness and the production of immutable mobiles, fieldwork is not compromised if it discovers something that exists only in one place. Kinsey's educated intelligence granted taxonomy an ontologically priority over other human sciences, gerrymandering his survey as the framework within which other smaller studies of human sexual behavior could be located and localized. Only taxonomy "provides an over-all survey" while "the experimental techniques are better suited to the examination of ultimate details" (*SBHM*, 18). However, Terman's critique threatened to unseat Kinsey's taxonomy from this central normative position. As with his earlier gall wasp taxonomies, Kinsey's claims to have discovered new forms of natural diversity in human sexuality based on self-reported histories that were derived from "proving the answer" could be interpreted as a discovery of important natural variation or as a constructed artifact of his own too fertile imagination.

Of course, unlike Terman, many people *were* quite convinced by the trustworthiness of Kinsey's published work. According to Sarah Igo (2007, 236), Kinsey's work both increased trust in aggregate surveys and enthusiasm to participate in them. Latour's theory of scientific inscriptions suggests one reason for such trust which points to the third influence of Pearl on Kinsey. *SBHM* is full of graphs. These "accumulative incidence curves" are variations on actuarial ogives (see fig. 4). They represent the proportion of people who *did* report a certain experience by a certain age to the interviewers, compared to the same proportion that *could* have reported such an experience. Latour (1990) argues that graphs deceive us modern folk into thinking that we are looking directly at natural phenomena even though we are really looking at human-made Euclidean *representations* of those phenomena constructed for rhetorical ends. Because graphs appear not to be representations at all, Latour imagines that they "enroll allies" in scientists' contests about what reality might be, leading scientists to proliferate graphs in modern scientific cultures which represent competing versions of what reality really is. Latour's thesis is certainly relevant to Kinsey's rhetoric, particularly those points where he noted that the "smooth trends" of accumulative incidence curves argue for natural laws of variation and should diminish readers' concerns about their contamination by "the vagaries of memory" (*SBHM*, 132).

Graphs can also effect power by being objects of mystery as well as trans-

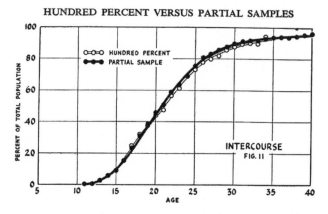

Fig. 4. Sample Accumulative Incidence Curve. *SBHM,* 101.

parent representations, and Kinsey's graphs conceal as well as display. Many readers will struggle to make sense of the sexual history displayed in figure 3 without access to Kinsey's codes (*SBHM,* 72). But many others may also wonder what meanings to draw out of the accumulative incidence curve shown in figure 4. Kinsey presented a detailed explanation of the calculations that lay behind the accumulative incidence curves (114–17), but his idiosyncratic quantitative logic is not easy to follow. Indeed, the accumulative incidence curves illustrate the arguments of anthropologist Wolff-Michael Roth (2003) that graphing conventions are intelligible only within particular scientific subcultures, whose rules of inference must be learned and managed. The graphs do not make men's sexual "nature" a matter of science by making clear natural laws, but rather make differences between men something to wonder at with some puzzlement.

Indeed, the fact that sexual behavior was being graphed may have been more salient to many readers than the information about sexual behavior within the graph itself. Graphing may also do different things than convince when applied to sex. Both in Kinsey's time and in ours the practice of tabulating and graphing sexual behavior has appeared to many people to be so devoid of a humanist, social, or emotional characteristic that it prompts the recognition that this representation of human sexuality must be incomplete in some crucial respects.[15] Anticipating such resistance, Kinsey described "a widespread feeling that statistics are cold" and "cannot measure human emotions which, after all, are involved in all sexual activities as well as in many other human problems" (*SBHM,* 20). The unsatisfying nature of cold graphs as visual metaphors for human sexuality has been linked to the constructedness of Kinsey's science in popular culture as well. Cole Porter wrote Kinsey's name into the song "too darn hot" for the 1948 musical *Kiss Me Kate*, but the use of Kinsey's name was omitted from the subsequent film of that musical. In Bill Condon's 2004 *Kinsey* film, the original version of the song appears.[16] As the lyric is heard "and when the thermometer goes way up and the weather is sizzling hot . . ." on screen a blank accumulative incidence curve appears and fills up with data points.

Finally, graphs can also open up science to scrutiny. Kinsey admitted that "one of the questions most frequently raised about the present research, and a thoroughly legitimate question about any research, concerns the possibility of another investigator duplicating the results" (*SBHM*, 133). To address this issue, *SBHM* reported similarities in the data resulting from interviews conducted by different interviewers, in two pages of text and eight pages of tables and graphs (133–43). But the graphs did not enroll Terman as an ally. He scrutinized the data carefully and spotted differences between interviewers' data, splitting Kinsey's intelligence off from that of his assistants, and suggested that duplication of the results had not even been achieved by the original research team (Terman 1948, 444). Scientists are not so robustly or routinely duped by the realism of particular graphs as Latour might suggest.[17]

Gentlemen and Horse Traders

If Kinsey's graphs confused and shocked as much as they convinced, then perhaps trust and distrust in his science might have depended upon some other more complicated social psychological dynamic. Informed by the sociological turn in science studies, the historian Stephen Shapin has argued that the dynamics of face-to-face trust in small human communities are much more pivotal to the apparent "hardness" of modern science than is commonly imagined. Shapin describes how early modern scientists were presented with a dilemma of trust that also confronted Kinsey; how can one engender trust in empirical matters that cannot be directly observed by everyone? Shapin (1995) argues that, as a partial solution to this problem, early modern scientists, particularly Robert Boyle, adapted codes for chivalrous gentlemanly conduct to craft the persona of the new experimental philosopher, who was uniquely suited to speak directly for nature.

Moreover, this history remains living; Shapin argues that throughout the twentieth century the knowledge taken to be objective remained "underwritten by the participation of 'gentlemen, free and unconfirm'd,'" rather than by "institutions which most vigilantly constrain the free action of their members" (1995, 413).[18] The social psychology of trust in Kinsey's science— the most popular and impactful science of its own time—then provides a test of Shapin's thesis. Trust and distrust in Kinsey's science can be explained by attending to three particulars that Shapin describes as being brought about by the gentlemanly persona of early modern scientists.

First, Boyle's persona depended upon his financial independence, which had been afforded by his father's earlier exploitation of Irish peasants. Men who had to earn a living from gerrymandering the value of natural things were considered too easily corrupted to be trusted to speak for nature impartially. Kinsey's sample included many categories of ne'er-do-wells which the category of "gentleman" might be defined against. The trustworthiness of Kinsey's claims rested on the trustworthiness of such men's voluntary accounts of themselves. Of course some of Terman's studies, notably *Sex and Personality*, also included diverse samples, but Terman treated such men as passive objects of a scrutinizing science; he did not trust them to author accounts of themselves.

The contrast between Kinsey and Terman in this regard is most obvious in the engagement of incarcerated men convicted of sex offenses. To complete the M-F studies in San Quentin prison, "Dr. Kelly followed the rule of having someone else present with him whenever he interviewed one of the homos to avoid blackmail attempts."[19] In contrast, Kinsey traveled to the prison himself along with his assistants, banned guards from his interviews, and put blankets on the wall to soundproof the interviewing room.[20] While Kelly and Terman used the records of prison officials, and the gossip of fellow prisoners, to justify their conclusion about the reality of the incarcerated men's homosexuality, Kinsey was careful to insulate prisoner's reputations from the effects of such intelligences. He and his interviewers refused to wear safety whistles when alone with the prisoners. Kinsey assumed that admission of homosexuality within the prison could lead to such penalties as "severe corporal punishment, loss of privileges, solitary confinement, and often an extension for a year or more of the sentence of an inmate who is discovered, suspected, or merely accused of homosexual relations" (*SBHM*, 129). Against such threats and ever-present gossip among fellow prisoners, "to persuade such an inmate to contribute a record of his activity while he is still in prison, is a considerable test of the ability of an interviewer" (129).

Second, Robert Boyle described the particular "Heroicke spirits" of those men who, by nature or by mental discipline, were best suited to natural philosophy (Shapin 1995, 158–62). Boyle described trustworthy gentlemen as those whose senses could perceive the world accurately, immune to the many "idols" that could distort a scientist's representation of nature. Such men had unique intellectual capacities that allowed them to integrate those perceptions directly, capacities which women and vulgars lacked (1995, 77). Kinsey's trust in his participants was limited by his evaluations of their Intelligence. It was impossible to derive sexual histories from people with IQ scores lower than 50, and for those whose IQ lay in the 50–70 range, "interviewing becomes a slow process in which each idea must be given plenty of time to penetrate, with endless repetition, and with a vocabulary which is confined to the simplest of words, both in the sexual vernacular and in the references to commonplace activities" (*SBHM*, 50). IQ defined one limit of the group who were left behind by the "educated intelligence" of his sexual science.

But Kinsey insisted upon new forms of "educated intelligence," new mental and social dispositions for scientists that did not yet exist. He expressed a worry about the capacities of upwardly mobile college-bound boys to engage in sex research. As Kinsey's data suggested that college-educated men had had particular kinds of sexual experiences, "the college-bred interviewer needs to go to considerable pains to limit his vocabulary to the relatively few words that are employed by persons in lower educational levels" (52). The "educated intelligence" that Kinsey hoped his study would bring about did not sit neatly within the existing structure of the university, but required an imagination of a future form of research training that had not yet come into being:

> Learning how to meet people of all ranks and levels, establishing rapport, sympathetically comprehending the significances of things as others view them, learning to accept their attitudes and activities without moral, social, or esthetic evaluation, being interested in people as they are and not as someone else would have them, learning to see the reasonable bases of what at first glance may appear to be most unreasonable behavior, developing a capacity to like all kinds of people and thus to win their esteem and cooperation—these are the elements to be mastered by one who would gather human statistics. When training in these things replaces or at least precedes some of the college courses on the mathematical treatment of data, we shall come nearer to having a science of human behavior. (*SBHM*, 35)

However much Kinsey's epistemology of sex research allowed the partiality of middle-class perspectives on sexuality to be glimpsed, he insisted that his interviews embody Audre Lorde's "mythical norm" in other respects. Kinsey insisted that interviewers would be men, to avoid concern about women who talked about sex; that all be married, to avoid concerns about licentiousness; and that none would be Jewish, Catholic, or Black—to avert any effects of prejudice toward such groups among research subjects. Like G. Stanley Hall, Kinsey didn't live up to the prescriptions of this own exclusionist ideology. Clyde Martin had neither a wife nor a Ph.D. when he began interviewing for Kinsey and, for a time, the Catholic psychologist Vincent Nowliss was a part of the interviewer team. Kinsey's particular trust in those groups from which gentlemen might be drawn was criticized by later feminists.[21] However, trust in such gentlemen remained an unspoken and untroubled norm in Kinsey's engagement with Terman.

Third, Shapin notes that Boyle's legitimacy required a *chaste* persona, celebrating Platonic love and the power of the spirit over the flesh. The chastity of the scientific imagination was also material to Terman's refusal to believe Kinsey's science. Specifically, Kinsey claimed that young men's future class trajectories were predictable from their early sexual experiences. Differences in rates of *nocturnal emissions* particularly stretched the limits of Terman's imagination of how education, class, and sexual behavior might be interconnected. Such differences pointed to sexual experiences that fit poorly with the definition of morality that Terman had used in *The Measurement of Intelligence*. Nocturnal emissions were an expression of sexuality that relied neither upon the ability to foresee the future or the capacity to exercise self-restraint. "Such differences in a type of sexual behavior that is non-volitional render suspect all of Kinsey's data on outlet as related to occupational mobility" (Terman 1948, 454).

Terman's skepticism of these data about class differences was not matched by all gentleman scientists of his time. Kinsey had made one possible link explicit which Terman had chosen not to engage:

> Nocturnal dreams may depend upon an imaginative capacity, in something of the same way that daytime eroticism is dependent upon the individual's capacity to project himself into a situation which is not a part of his immediate experience. It may be that the paucity of overt socio-sexual experience among upper level males accounts both for their daytime eroticism and for their nocturnal dreaming. (*SBHM*, 345)

Like Cowell's romantic account of his own creative mind, Kinsey's explanation of nocturnal dreams drew upon ideas about the power of the imagination. While Terman accepted the truth of Cowell's claims about his active imagination, he remained silent in response to Kinsey's explanation of why some boys might have capacities that allowed them both to conjure up erotic events and to climb the social ladder. Terman's failure to engage Kinsey did not go unnoticed by a third gentleman scientist. The economist Eli Ginzberg (1948) interpreted such class differences in sexual behavior in *SBHM* as the result of "discounting the future" to a greater extent among poor men than rich men, a hypothesis that had much in common with Terman's view that intelligence was the basis of morality. Ginzberg's interpretation of Kinsey's results appeared in print before Terman's review, and Terman sent Ginzberg a reprint of his own review upon its publication. Ginzberg replied with a generally positive letter to Terman, but suggested that he had been "a little harsh" in his criticism and incredulity, suggesting that his own theory of social class might help to solve the mystery of the nocturnal emission rates in Kinsey's data.[22]

Terman could clearly admit that academically promising working-class boys might defer sexual pleasure, but he did not engage this causal hypothesis about how the effects of their self-denial would leak out unconsciously. Such leaking might even have suggested a process of unconscious association, akin to the sort that led Hall to worry about the risk of smart boys masturbating, albeit an association that was repackaged as a gift—as an "imaginative capacity"—by Kinsey. Terman preferred to withdraw his trust from this supposed "educated intelligence" of human sexuality, and to keep his imagination chaste and his science clean of such matters. In other words, the Kinsey–Terman debate demonstrated how norms of disinterested, impartial, chaste gentlemanly conduct continued to shape both the dynamics of trust and the objects of knowledge in this new emerging sciences, very much as Shapin has suggested.

Kinsey's science both adhered to and flouted these conventions. As noted above, his science inculcated trust and rapport among a diverse group of people, even though his variable practices were likely to engender distrust and afforded a range of criticism. Robinson (1977, 49) described Kinsey as departing from norms of scientific conduct only by making sex an object of science. But the effects of Kinsey's science on the public went beyond that of distant disinterest in the objects of study. Even in *SBHM*, Kinsey engaged his readers not only as a scientist presenting findings. Rather unusually, he also coaxed his reader's trust in his science with the promise of greater understanding. A reader "will never understand this study until he comprehends the human drama that has been involved in securing the data" (*SBHM*, 36).

Like the upwardly mobile adolescent who experiences frequent nocturnal emissions, such trusting readers would need a kind of "educated intelligence" that allow them to imagine interview scenes where dirty talk has been exchanged, but which the readers could not have experienced themselves. On the very question of *why* readers should extend their imaginative capacities, Kinsey does not appeal to norms of gentlemanly conduct but draws on a very different metaphor:

> It has been asked how is it possible for an interviewer to know whether people are telling the truth, when they are boasting, when they are covering up, or when they are otherwise distorting the record. As well ask a horse trader how he knows when to close a bargain! The experienced interviewer knows when he has established a sufficient rapport to obtain an honest record, in the same way that the subject knows that he can give that honest record to the interviewer. (*SBHM*, 43)

In other words, Kinsey's success in spreading a new intelligence of sex depended upon readers entering into a conversation about a new way of understanding differences in sexual behavior with a sense of something that felt like a free choice. In extending the scientific method to a new object, Kinsey changed the nature of that method, as routinely happens in scientific networks.[23] In the closing section of this chapter, I want to compliment the question of *why* people trusted Kinsey by asking how people are moved as a consequence of extending such trust to talk about sex with scientists, particularly those scientists, who—like Kinsey—offered no explicit promise of individual redemption of the body, mind, or soul.

Hot Cognitions and the Dynamics of Initiation

Kinsey's note that statistics were "cold" was not the only time that he drew on the Galenic metaphor linking heat for human passion, or voiced the idea that sex research required "something more than cold objectivity . . . in dealing with human subjects" (*SBHM*, 42). How did people respond to these "warm" encounters? How did Kinsey's work change individuals' resting states? That work has left numerous accounts that cross class boundaries, and that bear witness to the fact that Kinsey's interview technique not only turned people into case histories on data sheets but also turned strangers into friends and doubters into allies. Many people became multiply positioned within the project both as researchers and as data points; all of Kinsey's associates contributed their histories to the project, as did Kinsey himself. For many people, the experience of being interviewed by Kinsey must have been unmoving, but many accounts suggest that the experience engendered trust in the larger project. Such was the case with Mrs. Nellie Showers Teter, the first woman trustee of Indiana University,[24] Robert Yerkes and George Corner from the CRPS,[25] gay men like Samuel Stewart who became unacknowledged collaborators,[26] and the committee of statisticians who ultimately audited the methodology of the Kinsey reports.[27] For still others, having one's sexual history recorded expanded the range of what could be discussed with others. "I found myself," wrote Wardell Pomeroy, "telling him things I had never dreamed of telling anyone else. Occasionally, as he deftly and persistently questioned me, I hesitated a moment, but then I said to myself, 'Of course, I must'" (1972, 98). The interview could transform intimate relationships; in anticipation of giving her sexual history to Kinsey, Martha Pomeroy exchanged hers with her husband.

Of course, the technique did not always work. Not everyone agrees to talk about their sex life with a researcher, and many people need to be coaxed by someone they personally trust before they will do so. For example, in March 1948, prior to writing his review, Terman wrote to Yerkes that he viewed Kinsey's work as "an enormously important contribution, but there are a number of things in it which I believe to be questionable."[28] Yerkes replied that Terman might think differently if he had been as personally involved in the project as he himself had been.[29] By relaying his personal experience to broker trust, Yerkes was behaving in a manner similar to several other people interviewed by Kinsey. One way that Kinsey hoped to avoid criticisms about the bias of his sample by recruiting "100 percent groups" in which *all* members of a preexisting social "group" were interviewed (*SBHM*, 39–40). The method of recruiting 100 percent groups relied on securing the trust and interest of "opinion leaders" within those groups who would be interviewed first and who would next broker the evidence of experience to create the conditions for consent to participate among other group members. Kinsey required opinion leaders most profoundly when working with those groups that were categorically excluded from acting as interviewers; such leaders mattered most with "poorly educated groups, and among such minority groups as rural populations, Negroes, segregated Jewish populations, homosexual groups, penal institutional inmates, the underworld, etc." (40).

How did Kinsey both draw upon and intensify the desire for a certainty about sex that Foucault called the "will to truth"? What did Kinsey draw out and transform in people through the experience of "subtle conversion of individual experience into social scientific data?"[30] Such spreading interest and trust in Kinsey's research is easier to psychologize from the present moment than from Kinsey's own. An experimental social psychology of the concept of "cognitive dissonance" began around the time of Kinsey's death. Developed by psychologist Leon Festinger, in part while he was on the faculty of the Stanford psychology department, the concept attempted to name an unpleasant state of embodiment caused by the recognition that one is holding inconsistent beliefs, or behaving in a way inconsistent with one's beliefs. Festinger's *A Theory of Cognitive Dissonance* (1957) was a pointed critique of behaviorism, the then-dominant school of thought in experimental psychology. Behaviorism had been ordered by an imperative *not* to speak of internal mental states such as desires and cognitions. Rather, behaviorists aimed to explain and predict patterns of human behavior as consequences of contingencies of reinforcement and punishment.[31] Because of their principled resistance to the reification of mental states—such as sexual desire and cognitive representation—Kinsey's taxonomic approach of sexuality, Foucault's rejection of sexual "desire" in favor of pleasure, and Latour's notion of the actor-network all share a family resemblance with behaviorism's aversion to speaking of "inner" psychological states.[32] Transgressing these norms, Festinger's theory of cognitive dissonance might help to both explain how people were moved by being interviewed by Kinsey and how Kinsey's own psychology might have accounted for the social psychological processes that occur when people talk about sex. Unsurprisingly, theories which suggest that cognitions are shaped by unconscious motivations are known among experimental psychologists as theories of "hot" cognition.[33]

Festinger and his students made cognitive dissonance an enduring object of study within social psychology for over half of a century. Using both observations and experiments, Festinger argued that people's investments in things, groups, and practices were not simply increased when rewarded or withdrawn when punished. For example, in one pivotal observational study, Festinger and his colleagues (1956) infiltrated a modern cult organized by belief in the end of the world and the group members' unique salvation by alien life forms. Members of the group appeared to this team of psychologists to be *more* invested in their leader's apocalyptic prophecies after they had obviously not been fulfilled. In an early experiment (Festinger and Carlsmith 1959), students received either a $1 or a $20 reward for convincing another person to take part in a boring experiment that they suffered through themselves. Although the $1 provided "insufficient justification" for their lie, the participants who received it described themselves as genuinely enjoying the boring experiment to a greater extent than those who had been paid off more generously. In the case of both the observational study and the laboratory experiment, the behaviorist claim that reward increases behavioral frequency while aversive punishment diminishes it appeared to have been refuted by the evidence. Both the cults and the experiments suggested the need to posit something within, something that *moved* people's attitudes when they found themselves behaving in a manner that was "dissonant" with those attitudes.[34]

Cognitive dissonance research might help to explain deepening investment in Galtonian science, such as Kinsey's, which seemed to enroll allies by pushing people to talk about sex through individual and group dynamics, and by forwarding the hope that buying in to such science will bring all who so invest to a better tomorrow. Two early cognitive dissonance experiments are particularly germane to understanding the social psychology of Kinsey's sex research. The first (Aronson and Mills 1959) was conducted by Elliot Aronson and Judson Mills at Stanford. Sixty-three women students volunteered for "a series of group discussions on the psychology of sex," and the male experimenters predicted that their attitudes to such talk would become more positive as a result of group initiation. Those experimenters made frank talk about sex for the sake of science an imperative for their participants in a manner that resembles the recruitment of 100 percent groups. They told each female participant that "if one or two people in a group do not participate as much as they usually do in group discussions because they are embarrassed about sex, the picture we get of the group discussion process is distorted" (178). The experimenters used their bogus concern to lead each student to believe that she would have to pass an "embarrassment test" before joining the discussion. Only one woman exited the experiment at this suggestion. Each of the remaining women were allocated to one of three conditions. In the "control" condition, there was no initiation. In the "severe" condition, the women had to read aloud "12 obscene words, e.g., fuck, cock, and screw" and "vivid descriptions of sexual scenes from contemporary novels." In the "mild" condition, women read words like "prostitute, virgin, and petting" (178). All participants then listened in on a prerecorded discussion on the sex lives of nonhuman animals. The discussion failed to titillate the male experimenters also, who described it in their report as "one of the most worthless and uninteresting discussions imaginable" (179). One further woman detected that the experiment had been a deception. Among the remainder, those in the severe initiation condition rated the discussion more positively than those in the other conditions. The authors concluded that the embarrassment of severe initiation had engendered "cognitive dissonance" between the behavior of talking frankly to pass the severe initiation and dislike of the group discussion. Having spoken frankly, the women who endured the severe initiation could only resolve their internal dissonance by rating the group discussion favorably.

More than Kinsey's obviously incomplete descriptions of why people took part in his sex survey, this experiment suggests a psychological processes that leads from talking about one's sex life to a scientist in surprisingly open terms to a more positive attitude towards a science that is built up from such conversations. I am not arguing that the cognitive dissonance tradition can sort out the wheat from the chaff of Kinsey's science in any absolute sense. Rather, it seems to me that "cognitive dissonance" constitutes a useful metaphor for conceptualizing the "positive unconscious" (Foucault 1973, xi) of Kinsey's science—that which Kinsey did, but which his behaviorist psychology inhibited him from encoding as knowledge.

However, I am cautious of suggesting this social psychological explanation without also placing the sexual politics of the scientific subculture of "cognitive dissonance" in the foreground as I go. While Aronson and Mills's experiment became a frequently cited textbook classic, the experiment has also been critiqued in recent decades because of the flippant style that the experimenters adopted toward the women in their study—an attitude that Kinsey trained his interviewers to avoid.[35] Experiments with different sexual politics were more quickly forgotten and subjected to critique because of the possible harm involved in changing research participants' relationships to their own sexualities. Under Festinger's supervision, psychologist Dana Bramel conducted experiments in which he emphasized to male college students the unconscious nature of sexual arousal. Some were given false biofeedback suggesting that they were responding erotically to "handsome men in states of undress" (Bramel 1963, 319). All participants then read what they believed were genuine responses of other male research subjects to a projective test, and all were asked to rate the extent to which those responses indicated "homosexual tendencies" of those subjects. When the bogus subject-who-had-taken-the-projective-test had been described as another male student, participants who had been deceived into thinking that they themselves had earlier responded in an erotic way to the handsome, scantily clad men attributed more homosexual tendencies to those subjects who had not been so deceived. Similarly, these participants estimated the proportion of homosexuals in the general population to be greater than did participants in other conditions. Bramel concluded that the biofeedback caused cognitive dissonance, resulting in attributions of "homosexual tendencies" to others in the interests of rationalizing one's own. Unlike Aronson and Mills's experiment, Bramel's was rapidly critiqued as unethical because it was considered to have been *too* psychologically effective.[36] The cognitive dissonance tradition is not conscious of the ways that heteronormativity have shaped it. Deception is a dubious but common practice in social psychological research. A deception experiment in which women are coerced to talk about sex has become a textbook classic. A deception experiment in which young men were led to think they experienced homosexual desire has been written off as unethical.

I think that the tradition of studying cognitive dissonance has paid a price for this unthinking heteronormativity, as Bramel's experiment has considerable explanatory power. In that experiment, participants projected homosexuality onto others as a "defensive projection" of their own apparently homosexual interest. Similarly, Terman and Oden's citation (1947) of Kinsey's and Davis's (1929) sex surveys projected homosexuality onto the general population to defend the normalization of the gifted. Bramel's experiment contained an important further manipulation. In Bramel's experiment, some men who were given the false feedback were given an opportunity to impute homosexual desires to other college men, while others were given an opportunity to impute those desires to a resident of a local jail. Only in the former condition of the experiment did cognitive dissonance appear to be resolved by the projection of homosexuality onto other men. Like Terman and Kelly, the smart college men in Bramel's experiment marked lines of inclusion and exclusion in their gender identity with universities and prisons. As such, Bramel's results accord with Kinsey's insight that college education ill-equips men to listen, without judgment, to the sexual histories of men whose life histories were very different than their own.

Finally, there are good reasons for skepticism about *my* particular use of cognitive dissonance as a metaphor for understanding the dynamics of power in Kinsey's science. After all, I was initiated through the rituals of that same psychology department of Terman, Festinger, and Bramel, and accepted the invitation to engage in the "human drama" of social psychology's tradition of deception-based experiments. However, I am very far from being the first to use psychological techniques as metaphors for Kinsey's science, and would suggest that cognitive dissonance might have some advantages over alternatives. Psychological metaphors and explanations are already at play. For example, Miriam Reumann (2005, 13) analogized Kinsey's science to the

Rorschach test, on the grounds that Kinsey's data allow any number of stories to be projected onto them. Into the twenty-first century, projective tests have remained a widely used metaphor for the uncertainties and pleasures of psychological interpretation even as the claims of their advocates have been repeatedly embarrassed by scientific studies that contradict them.[37] But Bramel's use of projective tests within his experiment locates his work in the Cold War period when confidence in the power of such tests to discern a man's sexuality was at its zenith. By the time that Bramel's experiment was published, experimental psychologists had reason to doubt that Rorschach experts could detect hidden homosexuals and were increasingly using the test to reflect back on the subjectivities of the clinicians who used it.[38]

Bramel's hypothesis that cognitive dissonance is relieved by the knowledge that one's deviant desires are shared by others is closer to Igo's claim that the Kinsey studies helped to relieve people from a sense of individual guilt about their sexual practices. Like Kinsey's science, and unlike familiar metaphors about the subjectivity of Rorschach interpretation, Bramel's experiment was premised on the knowledge that being involved in science did things to people, rather than simply representing them. We will continue to consider the range of imagination as we shift from defensive projections about sexual practices among the present population to projections about collective histories in the next chapter.

6

Ancient Ascetics and Modern Non-Americans

As long as Christians saw themselves in some way the extension and fulfill-
ment of Jews, they needed Jews in their conceptual framework.

Sander Gilman, *Freud, Race, and Gender* (1993, 9)

In this chapter, I will discuss the last of the four criticisms that Terman leveled against *Sexual Behavior in the Human Male*, a criticism which centered
on the sense that Kinsey made of differences between secular and religious
men. Much Kinsey scholarship—including this book—has implicitly located
Kinsey as a forerunner to new left movements such as second wave feminism,
the lesbian/gay rights movement, and other transformations of the "sexual
revolution." Terman's critiques show differences from Kinsey with regard to
the moral/intellectual value attached to the precocious emergence of male
sexual behavior and its orientation toward the pleasure of the self, or of other
women and other men. Their debate could suggest that Kinsey was more
"inclusive" and Terman more "exclusionary"; that Kinsey was more "liberal"
and Terman more "conservative"; or even that Kinsey was more "modern"
and Terman more "traditional." For example, Paul Robinson (1977, 50) has
identified the fundamental tenet of Kinsey's sexual ideology as *tolerance*.
However, tolerance found its exception in Kinsey's failure to appreciate the
importance of religion, and Kinsey's thoughts on religion in general, and on
Jewish men's sexual character in particular, trouble this characterization of
him as "tolerant." Rather, Kinsey's particular form of a "nonjudgmental" attitude to sex led him to conclusions that "othered" Jewish men repeatedly. If

we are not to locate Kinsey and Terman on a liberal-conservative axis, how then can we characterize the differences between these two "gentlemen"?

Historian of psychology Graham Richards (2010, 10) has recently argued that claims about "secularization" in modern societies represent a desire on the part of the secular human sciences rather than a firm empirical conclusion derived from the evidence of those sciences. Richards (2010, 3) bemoaned particularly how Foucaultian critiques of psychology have failed to attend to the normativity of Christianity—even as "Foucaultian discourse analysis" has come to name the analysis of dynamics of seemingly structural forms of power other than religion in psychological discourse, such as patriarchy, racism, and neo-liberalism, for example. A Foucaultian perspective would, of course, caution against the claims that Kinsey's work led to a liberation from repressive power. Foucault rejected narratives of progress based on the politics of liberation and ontologies of the nature of human sexuality. For Foucault, there has been no "historical rupture between the age of repression and the critical analysis of repression" (1978, 10), presumably even in Kinsey's work. I share Foucault's very skeptical orientation to discourses about sexual nature, and will give reasons to be more skeptical of Kinsey's account of sexual variation in this chapter. However, there are also reasons to consider Kinsey and Foucault's accounts of the history of sexuality as similarly invested in narratives of historical rupture that shape all resulting discourses in the "Western" scientia sexualis. In contrast, Richards urged caution about assuming that such transition points between religious and secular authority are ever clean breaks.

Not all areas of the history of the psychological sciences have neglected the history of religion to the same extent. Richards further noted that the history of psychoanalysis has more often attended to questions about religious influence than has the history of psychology. As we have seen in previous chapters, Terman's thoughts about the intellect were in accord with Freud's to a surprising degree, particularly with regard to the promise of gifted boys, such as Henry Cowell, "sublimating" their queerness into works of genius. Henry Cowell's life also points to the merit of Richards' suspicion that postmodern theory mismeasures both the extent to which and the reasons why religion has remained of determinative importance to so many people. Terman's secular psychology of gifted children has failed to ensure absolutely that gifted children were normal people, beyond suspicion; as I argue in chapter 8, very smart people are still sometimes regarded as dubious. Moreover, religious authority sometimes succeeds in limiting normalization where secular argument fails. For example, the arguments of the counselors of the Temple of the People successfully saved the younger Henry Cowell from incarceration for his homosexual behavior, while Terman's later more secular argument about sublimation did not save the older Cowell from a term in San Quentin. Of course, there are many more differences between these two attempts to protect Cowell from homophobic law, but they jointly remind us that being designated as "gifted" by a secular psychologist can afford less protection than inclusion in a modern religious collective.

Kinsey's account of sexual nature was not only in tension with Terman's over the sexual character of Jewish men, it was also in tension with psychoanalysis. Psychoanalysis provides a contrasting case to the values within Kinsey's work, including the limits of his tolerance that Jewish men lay beyond. In 1976, Paul Robinson stated that "Kinsey has had probably a greater influence on modern sexual consciousness than any other thinker since Freud" (115). The publication of Foucault's *History of Sexuality* shortly thereafter might render that claim somewhat old fashioned. Foucault's work effected a radical break of its own, shifting scholars in diverse fields away from a narrative of sex's modernization and liberalization to a history of "the West" shaped by the production of discourse. However, the contrast between Kinsey's and Freud's accounts of the history of religion suggest one way to query Foucault's general hypothesis that "it is in the confessional that truth and sex are joined" (1978, 61). As I described in the previous chapter, Kinsey's science certainly moved people, but it was not presented as a model of individual salvation, akin to either a talking cure or a confessional. The differences between Kinsey and Freud, particularly in their accounts of ancient religions, also inform Richards' argument that psychology continually strives—and fails—to adequately address fundamental questions of *value* that satisfy people in the way that religion so often does. Perhaps both the modernization of sex and its postmodernization after Foucault are more indebted to the religious traditions that they aim to move beyond than they currently appear to be.

Kinsey's Psychology of Religion

Kinsey's biographers have made no secret of the fact of his own negative attitude to the Christian religion within which he was raised. Each of the book-length biographies narrates how Kinsey's strict religious father played an important role in his original observance and later rejection of organized religion. Since Bill Condon's *Kinsey* (2004), many more Americans have become aware of this narrative; indeed Condon's film opens with scenes of Kinsey's father warning of the dangers of immorality risked by zippers in trousers and of the young Alfred praying to be relieved of the desire to masturbate. In December 1915, at the age of twenty-one while a student at Bowdoin College in Brunswick, Maine from 1914 to 1916, Kinsey joined a Congregationalist church and worked with the YMCA.[1] He stopped attending church at the age of twenty-six when he moved to Bloomington. For "cultural" reasons, Alfred Kinsey continued to send his children to Sunday school.[2]

While conducting sex research in the 1940s, a variety of religious actors and religious ideals frustrated Kinsey's efforts to develop a science of human sexuality. Threats to morality were the rationale for both local religious leaders and the Indianapolis physician Thurman Rice to approach Indiana University president Herman Wells about Kinsey's research. Wells then forced Kinsey to choose between teaching about sex and researching it. The research of psychologist Glenn Ramsey on the sex histories of schoolchildren in Peoria, Illinois, was similarly questioned by the president of the Board of Education there. Ramsey had begun his work independently of Kinsey, but his data were later incorporated into Kinsey's database, and Ramsey was briefly on Kinsey's payroll after Kinsey had secured CRPS funding. Ramsey attributed the scrutiny of his own research to one of his local opponents' "host of Catholic friends."[3] Much later, after hearing Kinsey lecture late in 1947, sociologist A. H. Hobbs wrote to insist that the moral obligations of scientists went beyond the collection of facts. Hobbs urged Kinsey to appreciate his responsibilities to moralize about abnormal sexual behavior to a much greater regard. Otherwise "you have violated an ethical code which many scholars have imposed on themselves in relation to the probable social effects of publicizing information regarding the 'Facts of life.'"[4] Like those experimental social psychologists who excluded Dana Bramel's experiment from their canon, Hobbs considered that a possible projection of large numbers of homosexuals onto the imagination of the national population transgressed the bounds of scientific gentlemanly conduct. On March 7, 1948, the New York Times carried an article by Hobbs, a précis of a more lengthy critique (Hobbs and Lambert 1948) that would appear in June in the American Journal of Psychiatry, aimed at protecting a suggestible public from investing trust in Kinsey's social project. Victorian opposition to sex research, grounded in Christian tradition, had not simply gone away. Rather, these examples of opposition to Kinsey's work demonstrate how understandings of sexual morality which appealed to Christianity for their authority partially shaped the debates about the sexual sciences that were coming into being at this moment. The continuing targeting of the transformations of the sexual revolution by the religious right in the United States, and the targeting of Kinsey's reputation in particular, shows that such processes are ongoing.[5]

Kinsey's decision to send his children to Sunday school for "cultural" reasons show an awareness of the modern processes by which secular and religious authorities have become more intertwined and harder to distinguish. For Kinsey himself, such intertwining rendered it difficult to tease out individual motivations for resistance to his science.[6] In *SBHM* (479), he voiced the idea that secular people who opposed sex out of seemingly secular "plain decency" were truly influenced by longstanding religious codes of morality. Kinsey's biographers have also braided the religious heritages of the West into their constructions of the deep motives of this most secular of scientists. Gathorne-Hardy (1998, 21) described Kinsey as battling the legacy of St. Augustine, for whom original sin was knowledge about sex. For Christiansen, Kinsey's scientific work "was imbued with a Puritanical ethic of work that gave it a value comparable to the Christian emphasis on faith," while "music was in a certain sense a religion with Kinsey" (1971, 176). Pomeroy similarly describes Kinsey's taste in music as "Catholic," "as in all the arts . . . his tastes were on the whole conservative" (1972, 89). Such analogies between diligent religious observance, conservative taste, and scientific discipline suggest that on the disavowal of passion is a key element of religious life. The association of religion with mastery of the passions by reason is also consistent with Kinsey's own psychology of religion, which centers particularly on the culture and character of Jewish men.

In *SBHM*, Jewish American men were singled out as strange atavistic creatures who enjoy little sexual freedom themselves, and who come to curtail the freedoms of others in the interests of a quite peculiar from of rationality that is easily made into a spectacle. If "sexuality" works through "disciplinary power" such that dominating agency lies with the one who asks the questions rather than the one who answers, then that power put Jews, more than Christians, in their place. Kinsey's taxonomy contained only three groups of believers; Catholics, Protestants, and Jews, but only the last of these three groups was linked to ancient texts in particular ways. The entry for "Jewish sex patterns" in the index to *SBHM* (797) directed readers both to the pages of the chapter in which the sex patterns of religious men were compared, and to "See also *Bible, Talmud.*" The equivalent entries in the index (797) about Catholic and Protestant sex patterns directed the reader only to the chapter's pages and not to any particular religious texts. In *SBHM*, living Jewish American men's sexualities were uniquely indexed to ancient texts in ways that Christian American men's sexualities were not.

This selective indexing was repeated throughout the text. While Foucault rejected the notion of a historical break in modern times between an era of repression and an era of its critique, he posited (1978, 58–73) a much earlier break that had led the modern Occident to develop along a particular path that had led to the construction of a *scientia sexualis*, which differentiated it from many Asian societies organized by an *ars erotica*. Kinsey similarly posited a historical break in the ancient world and described the Talmud as the source of a repressive influence on ancient cultures, including early Christianity. As Robinson put it, Kinsey "seemed to think that Jewish asceticism had in the modern world become the more or less exclusive property of Englishmen and Americans. Continental Europe remained in his imagination an almost Polynesian sexual paradise" (84). On the first page of the chapter on "Religious Background and Sexual Outlet," Kinsey asserted that "our particular systems certainly go back to the Old Testament philosophy on which the Talmud is based, and which is the philosophy of those Jews who first followed the Christian faith" (*SBHM*, 465). The chapter on "Marital Status and Sexual Outlet" similarly opened by distinguishing between "hedonic" interpretations of sex as used for pleasure and interpretations of sex as justified only by reproduction. "The Hebrews were among the Asiatics who held this ascetic approach to sex" (263), and Christian philosophy and American law were later derived from a confluence of Hebrew, Greek, and Roman influences. "Many of our present-day attitudes on sex are matters which were settled in the religious philosophy of the authors of the Old Testament and even among more ancient peoples" (415). The chapter on "Pre-Marital Intercourse" noted that modern prohibitions against premarital sex were formed, in part, "from the tremendous importance which is attached in Jewish codes to the virginity of the female at the time of marriage" (547). In this book on the sexuality of men, references to the particulars of women's bodies were rare. This chapter made an exceptional spectacle of concern with virginity and women's bodies: "First generation immigrants in some parts of this country today may still send the blood-stained napkin back to relatives in Europe" (548).

Kinsey assumed that this ancient influence had a particular effect on modern Jewish men and invoked this influence to explain various group differences. In *SBHM*, religious men were described as less sexually active than other men, both in general (469) and within marriage (482). Moreover, men categorized as Orthodox Jews were less sexually active than those categorized as devout Catholic or Protestant Christians (469). Kinsey described "the sexual philosophy of the Jews" as one which conflated sexuality and reproduction, and which acted as a "considerable impetus" to the taboo against masturbation in the Christian religion (473). Kinsey insisted that "the Jewish boy is definitely affected by the old-time Hebraic law on this point" (473). "The often lower rates of the religiously inactive Jewish boys indicate that even they are not entirely free of the ancestral codes" (476).

Kinsey clearly thought that such influences of religion were "unnatural"—in the pejorative sense, and that nature would out: "it is clear that there is a sexual drive which cannot be set aside for any large portion of the population by any sort of social convention" (*SBHM*, 269). Terman spotted how Kinsey's own data argued against this claim because they "suggest that religious attitudes have a considerable influence on most types of sexual activity" (1948, 455). Terman was also unhappy with the quality of that data, pointing out "the small N's of Catholics and Jews" in Kinsey's sample (455). Indeed, as Terman pointed out, the sample contained "only 59 Orthodox Jews in the entire U.S., all of the college level!" (455).

However, Kinsey's assumption that this form of originally Jewish repression had been successfully secularized also made available an explanation of group similarities. Kinsey found homosexual sex to be equally common among devout Catholic, Protestant, and Jewish men. However, he still explained how "the Hebrews, in contrast to some of their neighbors, attached a severe religious condemnation to homosexual activity. . . . There has, in consequence, been a continuous history of condemnation of the homosexual in the Christian church from its very beginning" (*SBHM*, 483). Consistent with Kinsey's location of homosexual men on the Western frontier, he described the European Jewish sexual philosophy as other to the "educated intelligence" that his own American sexology could promise to bring about. Present-day attitudes reflected "matters which were settled in the religious philosophy of the authors of the Old Testament and even among more ancient peoples," but "scientific analyses will quickly modify such deep-rooted behavior" (415). Like the European gall wasps whose migration had inhibited the proliferation of species on American soil, a Jewish culture, European in its origins, was positioned as a brake on the evolution of distinct forms of American life, including Kinsey's own science.

Kinsey's cultural essentialism was exemplary of "othering"; he admitted of more than one species of Christian but assumed that Jewish identity was monolithic. Jewish men were located outside the norm for the American nation state, constituted its defining problematic, and provided the explanation of both difference and similarity within it. Robinson notes that this commitment to naturalism drew fire from Kinsey's humanistic critics (56). Most crucially, Lionel Trilling's widely read review of *SBHM* cogently describe how "othering" was made inevitable by Kinsey's commitment to both Darwinism and empiricism. Trilling noted that while "the Report directs the harshest language toward the idea of the Normal," it was "by no means averse to letting the idea of the Natural develop quietly into the idea of the Normal" (1954, 222). The taxonomist's emphasis on variation and his Galtonian normalization of early, varied, and prolific expressions of male sexuality inevitably meant that men who were less sexually precocious and expressive were less "natural," and less ideal than others. American Jews were not quite "natural" by Kinsey's terms; they were too closely indexed to "culture," albeit quite ancient forms. However, Trilling's critique of *SBHM*—unlike Terman's—did not explicitly note that it was religious men, particularly Orthodox Jews, who were made the "other" of this scheme.

Kinsey's account of the sexual psychology of the Jews was not entirely atypical for his time. Published between the two Kinsey studies, the influential ethnography of shtetl life in Eastern Europe, Mark Zborowski and Elizabeth Herzog's *Life is with People* (1952), popularized an essentialized account of Jewish Americans. Based on interviews with European-born Jews then living in the United States, *Life is with People* defined "traditional Jewish culture" for a generation as uniformly oriented towards promoting the life of the book and the intellectuality of sons.[7] However, there were also alternative cultural essentialisms available. A very different notion of Jewish sexual character was drawn into psychologist David Bakan's account of the influence of Jewish thought on Sigmund Freud, *Sigmund Freud and the Jewish Mystical Tradition* (1958). Richards (1997, 126) describes Bakan's book as "one of the most enduring works" in the history of psychoanalysis. Therein Bakan concluded that the Kabbalistic tradition led to the emphasis in psychoanalysis on sex as the secret source of all things, insisting even that "never in the Jewish tradition, was sexual asceticism made a religious value" (1958, 272). As such, Freud's understanding of Jewish sexual character—or at least Bakan's account of it—is in direct contrast with Kinsey's assumptions, which made living Jewish culture the "other" to sexual nature in *SBHM*.

Bakan was not the only psychologist of his time who was influenced by Freud or whose work contrasts with the unconscious othering that occurs in Kinsey's work. Psychologist Gordon Allport understood how Jews could be othered both by association with sexual exuberance, as in Bakan's work, and

with sexual asceticism, as in Kinsey's. Moreover, Allport drew on Freud's notion of *projection* to do so. Psychological studies of ethnic and racial prejudice
proliferated in the postwar period in the United States, reversing and mod-
ernizing the theories of "race difference" more common in earlier decades.[8]
As ethnic stereotypes about many European groups declined, Blacks and
Jews loomed largest in this literature as the targets of "prejudice."[9] Drawing this work together in *The Nature of Prejudice* (1954), Allport noted that
stereotypes often take the form of projecting unwanted characteristics onto
others. Within the United States, "the Negro reflects our own 'id' impulses;
the Jew reflects our own violations of our 'superego.'" Allport continued that
in Europe "where there is no Negro minority, it is the *Jew* who is blamed for
the lechery, filth, and violence" (1954, 194). Kinsey had also noted a difference in the stereotypes about Jews in Europe and the United States but had
gerrymandered the difference to position his own account as unprejudiced.
He noted only casually that European anti-Semitism had led to the Shoah
when he described his own empirical conclusions about Orthodox Jewish
men's sexuality as "especially interesting, in view of the diametrically opposite opinion which recently stirred a considerable portion of Europe against
the Jews as a race" (*SBHM*, 485). The "prejudice" in Kinsey's text might be
called "modern" by twenty-first-century social psychologists, because it took
pains to define itself *against* prejudice.[10] However, many European Jews who
had their immigration to the United States blocked after the 1924 act—that
Kinsey had earlier praised as good taxonomic thinking—met their end in
the Holocaust.[11]

Trilling, Bakan, and Allport all used Freud to understand the dynamics
of othering in Kinsey and beyond. In *SBHM*, Kinsey had credited Freud
with the idea that sex is a biological function (263), and had granted some
validity to the practice of dream interpretation (525–27). Kinsey had also
taken issue with Freud's notion of pre-genital stages of psychosexual development (163, 180), his definition of masturbation (499), and his use
of clinical case histories to make claims about average patterns of sexual
behavior (34). Kinsey also described "sublimation'—that concept which
linked sexuality and the intellect most obviously—as an outgrowth of "ancient asceticism" (206), reading psychoanalysis as one of those forms of
seemingly secular morality that was nonetheless animated by ancient religious feelings.

The difference between Kinsey and Freud's approaches to individual
normalization was not lost on Kinsey's critics. Trilling particularly noted
that Kinsey had acknowledged his work's "debt to Freud" (1954, 213) but

critiqued Kinsey nonetheless for being insufficiently Freudian about such matters as conflating frequent sex with good sex, and forgetting that masturbation in children could be a sign of compulsion (220), and that among the "concepts of sexual aberrancy" homosexuality remains the "the most complex and most important in our cultural life" (227). In particular, Trilling's criticisms all point to Kinsey's shift away from the normalizing practices of confession, in which "sex" is made to signify individual character and individuals are specified on the basis of their sexual conduct. However, as Sarah Igo has noted, Trilling also valued Kinsey's work as culturally therapeutic.[12] Trilling described how "modern society has atomized itself" and "censored the folk knowledge of the most primal things." Trilling recognized that *SBHM* had created a collective consciousness that, by "establishing the *community* of sexuality" (1954, 213), was potentially culturally therapeutic. While Robinson similarly concluded that Kinsey's work "was fully conscious of the demystifying motif in his work" (118), individual comments suggest that some people found meaning that went beyond their individual selves by contributing to Kinsey's science. For example, Igo quotes one woman who was fascinated by the experience of giving her sex history while also remaining anonymous: "No longer was I an individual; only a statistic from a pattern along with thousands of other cases. I felt about as significant in the scale of life as a star a million miles away" (Igo 2007, 270).

While Kinsey's work was presented as empiricism of the most demystifying sort, both this comment and Trilling's reference to the primal and the collective suggests that the transformative effects of the Kinsey studies—which may or may not have been "therapeutic"—were anything but individual in their dynamics. Kinsey's sex survey was not a confessional. Indeed, as we turn to Kinsey's understandings of the dynamics of "ancient asceticism," we can see that his work was not only fully conscious of its demystifying effects; it was also unconscious of the dynamics of remystification involved in translating lives into scientific representations.

Confession and Initiation

Kinsey's partial understanding of the influence of ancient religion on modern life, encoded in his account of the influence of religious "asceticism," points again toward the conclusion that "initiation" and not "confession" might name the positive unconscious of his science. Genealogy often works by drawing together excluded and defunct knowledge, to trouble the truth effects of current hegemonic formations. "Initiation" is both excluded and defunct within Foucault's historical narrative about the confessional. On the one hand, Foucault insisted that the *scientia sexualis* of the Occident became differentiated from the *ars erotica* of older and less Western cultures with their "learned initiations into pleasure, with their technique and their mystery" (1978, 62). On the other hand, the *ars erotica* represent a kind of unconscious *within* the already unconscious modern dynamics of secular confessional practices. Foucault insists that Christian confession had much in common with an erotic art, including "guidance by the master along the path of initiation, the intensification of experiences extending down to their physical components, the optimization of effects by the discourses that accompany them." He also foregrounded the question of whether these dynamics of initiation continued to function in modern sciences. "And we must ask whether, since the nineteenth century, the *scientia sexualis*—under the guise of its recent positivism—has not functioned, at least to a certain extent, as an *ars erotica*" (70–71).

Foucault is most definitely shaming the secular advocates of *scientia sexualis* for their pretensions to have escaped the dynamics of older religious forms. However, it is less clear why "confession" should be primary and "initiation" only secondary in this characterization of the West. Rather, as the cognitive dissonance experiments described in the last chapter suggest, the investments in sex surveys accomplished through Kinsey's work may have proceeded through practices of initiation that were not confessional at all. The dynamics of initiation are also erased from Kinsey's history of religion. To support his understanding of the Ancients, Kinsey cited *two* histories of religion in *SBHM*; Geoffrey May's *Social Control of Sex Expression* (1931) and Samuel Angus's *The Mystery Religions and Christianity: A Study in the Religious Background of Early Christianity* (1925). May described how the Jews became distinguished from their neighbors by "control of human relations" (15) and traced an ascetic influence through to a critique of the laws of the twentieth-century United States; "the substantive law no longer expresses the existing attitude of the people as to private voluntary non-commercial aspects of sex" (212). May's linear account attributes a repressive influence in American sexual law to a direct influence of Christianity, which is itself attributable to the influence of Judaism.

However, the radical cleric Samuel Angus wrote a much more "contrapuntual" history of the early Christian church.[13] Kinsey cited his work selectively, excluding Angus's lengthy descriptions of the dynamics of initiation in the Mystery religions of the ancient world.[14] In Angus's account, Jews constituted the "main link between East and West" after their conquest by Alexander in 333 BC.[15] Thereafter, the Greek religion continued to contain unresolved tensions between Athenian and Dionysian impulses, quite in contrast to Kinsey's descriptions of the Ancient Greeks as singularly credulous of the idea that excessive sex and masturbation deprive the brain and spine of needed semen (*SBHM*, 297). While Angus described the syncretism of Jewish and Greek thought as one that fostered a climate in which ancient mystery religions began to thrive, Kinsey positioned Jewish religion, culture, and philosophy as the conduit of asceticism from East to West in the ancient world. But in Angus's narrative, the sexual culture of early Christianity had emergent properties that led to new kinds of investments in older religious rituals of initiation, troubling any attribution of twentieth-century state repression to events that occurred in the ancient world, or their influence, via Judaism and Christianity, on American law. State power over private subjecthood was not absolute or stable in Angus's telling of it. Rather, the mystery religions proliferated as Rome's imperial power over religion weakened. Centuries before Christianity became the religion of empire under Constantine, new temples were being dedicated to Isis during the reign of Caligula.

The mystery religions and their dynamics of "initiation" were excluded from Kinsey's account of the history of religion. Yet, perhaps "initiation" names aspects of the positive unconscious of Kinsey's science that "confession" does not. Like the demography of Raymond Pearl, the mystery religions that Angus described arose from the observation of birth and death in nature and were motivated by a desire to understand death's mysteries. Like the "educated intelligence" of Kinsey's science, initiates of mystery religions experienced rites of conversion that were scripted as secret. Much as Kinsey's science was visibly dubious to the social scientists of his time, Angus describes the mystery religions as "weak intellectually and theologically" (1925, 61). And yet, like those individuals included in Kinsey's inner circle, the initiates of mystery religions understood themselves to have transcended ordinary consciousness in an ecstasy that was beyond the understandings of ordinary philosophy; they were bringing about a more enlightened vision of the nature of sex than anything that had hitherto existed. The mystery religions often prescribed the orgiastic, an ideal that was as incompatible with Boyle's chaste vision of empiricism as Kinsey's invocation of the imaginative capacities of the upwardly mobile adolescent. But Kinsey's science was atypical in arguing for a particular education in sex research that went beyond the existing norms of his time, and it remains a black mark on his legacy that he filmed people having sex in his own home, still unable to square such practices with his public persona as a dispassionate scientist. If these analogies seem too strained, then it may be worth remembering that Angus himself urged readers to consider such analogies. He described those secular-religious organizations which Kinsey took as his targets as the vehicles through which to imagine the mystery religions of the ancient world. Angus conjured up an imaginary present in the 1920s in which "the Salvation Army, the Y.M.C.A. and the Masonic brotherhoods were to take away the prestige of the historic churches . . . and draw to themselves their membership" (1925, 235).

Matters of value and the history of power/knowledge are at stake in shifting from the metaphor of confession to the metaphor of initiation. Angus's history is dissonant with Kinsey's historical essentialism regarding the influence of monolithic Jewish "culture" and the resulting value of liberating people from repression of their varied sexual natures that that culture effects through American sex law. Indeed, if the mystery religions could arise and fail within the Roman Empire, and if such religions had counterparts in the modern United States, then it would be difficult to see how Kinsey's account of repressive state power over sexual nature would have retained its explanatory power. In other words, the exclusion of complicated narratives about polytheistic ancient cultures was constitutive of the account of natural sexual variation within *SBHM*, and the positioning of Jewish people as "others" to that nature.

Sublimation and the Sexual Politics of Smart Jews

While Kinsey's history of religion has rarely been interpreted as an expression of his secular WASPish background, Freud's final work *Moses and Monotheism* (1939) has often been read as a cipher of his relationship to Jewishness, and of the relationship of psychoanalysis to Jewish philosophy in particular.[16] I want to read it here as a contrasting case of how the dynamics of othering might be engaged in modernist writing about the sexual cultures of ancient peoples. At the center of *Moses and Monotheism* lies Freud's hypothesis that the biblical Moses was not Jewish but Egyptian, such that the invention of monotheism was an accomplishment that has been wrongly attributed to the Jewish people. Like the United States represented in *SBHM*, the Jewish nation narrated by Freud is a divided one. Freud describes Moses as a noble devotee of a monotheistic Egyptian religion whose Jewishness was an invention of the people whom he led out of Egypt. Freud's account then posits a second historical Moses, a priest of the Midianites from "the region South of

Palestine." The Egyptians and Midianites joined into one nation, which was divided between an intellectual (Egyptian) elite and a more sensual (Midianite) majority. Much like Kinsey's "upper-level" males, the Egyptians were a smaller but stronger group, and the Egyptians had learned from Moses "a more highly spiritualized notion of god, the idea of a single deity embracing the whole world, who was not less all-loving than all-powerful, who was averse to all ceremonial and magic, and set before men as their highest aim in life in truth and justice" (1939, 50). These Egyptians included the Levites, a "culturally superior" minority elite who transported history and writing out of Egypt and later formed the priestly class (37). Just as Kinsey described the middle class as exercising their moral authority through the rule of law in the United States, Freud described the laws of the Jews as an instance of repression enacted on the majority by the more educated, literate minority.

In other words, both Freud and Kinsey set themselves the task of narrating national histories in androcentric terms, and in which men were divided by power, intellectuality, and sensuality. Yet, these two wrote very different narratives with very different consequences. While Kinsey rejected "ancient asceticism" with its influence on Freud, in Freud's account the peoples who made up the Jewish nation initially varied in their asceticism. Freud's narrative was also invested in the cultural value of sublimation, in contrast to Kinsey's. In spite of recognizing the heterogeneous roots of the Jewish nation, Freud ultimately positioned monotheism and sublimation *as* valuable achievements. He described the Jews as being forced by historical circumstances to choose between the "cultivation of intellectual and physical activity" and declared that "their [the Jews'] decision [in favor of the intellect] was at least in favor of the worthier alternative" (115).[17] The value placed on sublimation in Freud's narrative could not be more different from Kinsey's critique of it as an unnecessary legacy of ancient culture.

Freud's *Moses* has had a lively time since Freud's death. To cut a very long story short, scholars have varied on the matter of whether Freud's *Moses* represents an expression of a problematic or useful ambivalence on Freud's part towards Jewish identity.[18] In recent decades, the Jewish historian Yosef Yerushalmi has used the discovery of a gift of an inscribed Hebrew Bible from Freud's father to Sigmund to argue that Freud privately possessed more educated intelligence about Jewish religion and culture than his persona allowed him to admit. In response, Jacques Derrida (1995) described Yerushalmi's work as using the authority of the archive to return Freud to an account of Jewish identity, selfhood, and memory which psychoanalysis strains against. The text has allowed a very wide range of interpretations of Freud's relationship to Jewish traditions to be elaborated. Most relevant to the "othering" of Jews within Kinsey's account of the United States is the appreciative reading offered by the postcolonial scholar Edward Said. Said voiced an admiration for Freud's *Moses*, written in the last years of the 1930s, for its reluctance to accept, for the Jews, the shelter of anything like an exclusionary "European" identity even in the darkest of historical moments. Freud allowed that "it is honor enough to the Jewish people that they could preserve such a tradition [monotheism] and produce men who gave it a voice—even though the initiative to it came from outside, from a great foreigner" (1939, 51). Said's reading of *Moses* is contestable; however, it is his *appreciative* mode of postcolonial critique that I wish to draw on here. Said set out not simply to critique European texts as hegemonic; but to demonstrate, in addition, how the texts that he critiqued might have enduring value beyond their own time and place. Said's conclusion that *Moses* might have value for times and places that Freud could not imagine might be confirmed by its contrast case with Kinsey's *SBHM*, a text that could too easily be seen as more "modern" than Freud's, less mysterious, or even as more fundamentally *tolerant* of difference.

These differences between Kinsey's and Freud's narratives about religion confirm Richards' idea that psychologists often aim their science at matters of value, but typically fail to address them in robust ways that compete with established religions. Terman's understanding of Jewish character and its implicit value judgments adds weight to this argument. A secularist himself, Terman wrote to Yerkes a year after *SBHM* was published that he had "lost all interest in organized religion when I was about 18 or 19 years old. This coincided with my enthusiasm for Darwin, Huxley and other modern scientists."[19] Terman's attention to Kinsey's claims about the effects of religion on sexual expression—and the lack of an adequate sample of Orthodox Jewish men—are akin to his other criticisms in that they focus on a moment where Kinsey appears—at least to Terman—to have gilded the lily of sex and to have underestimated the value of its substitution. However, Terman's science of IQ testing indexed Jewish people to "intelligence" in problematic ways. "Jewish" was treated as a correct synonym for "shrewdness" in one of the items on the original Stanford-Binet test,[20] and the children of Jewish immigrants to the United States showed up in disproportionate numbers in studies of gifted children—both in Terman's in California (Terman 1925) and in Leta Hollingworth's in New York (Hollingworth and Rust 1937). In an address to Sigma Xi at Stanford in 1941, Terman argued that "the Jewish child is under heavy pressure to succeed, with the result that he seems to accomplish more per unit of intelligence than do children of any other racial stock."[21] Gifted Jewish children grew up to be more educationally and financially successful than other gifted children, had "greater drive for vocational success," were more liberal in their political attitudes, and somewhat less likely to be divorced (Terman and Oden 1947, 310). While Kinsey's sexology tended to assume that Jews were largely unassimilated, Terman's differential psychology set up a double bind that many immigrant people will recognize; he romanticized an ideal of culturally distinct Jewish childrearing practices while also favoring assimilation over noticeable difference.

Terman's thoughts about the distinctiveness of Jewish intelligence still matter for us because they have been named in recent decades as the basis of an intellectual tradition that ontologizes Jews as possessing superior levels of inherited intelligence. Economist Miles Storfer had theorized that Jews, specifically Ashkenazi Jews, were high in heritable intelligence, and Storfer traced his intellectual roots to Terman.[22] Richard Herrnstein and Charles Murray's *The Bell Curve* (1994, 275) cited Storfer's work to argue that Ashkenazi Jews had particularly high intelligence for biological reasons. In response, historian Sander Gilman used this assertion to examine how the attribution of high intelligence to a group can be as damaging to their character as the attribution of low intelligence, such that the "thin ends [of the bell curve] define the problem—the 'normal' center is understood as the model of intelligence and therefore also as the model of virtue" (1996, 4). Gilman's hypothesis suggests that the normalization that occurs around intelligence tends to be Queteletian rather than Galtonian, a point that has merit because Jewish smarts were associated with sexuality abnormality. Gilman details how Jewish men's intelligence was also often seen as a compensation for a lack of physical manliness and strength such that the queerness of intellectual asceticism was particularly projected onto Jewish men. As such, theories such as Storfer's were "as likely to draw the virtue of the Jews into question as to posit them as a site of virtue" (6).

Storfer located himself within a tradition of thinking about the inheritance of intelligence that dated to Terman, and Terman's life and work show ambivalence in his attitude to the normality of religion. Like homosexuality, religion appeared as a sign of femininity in Terman's theory of gender. The masculinity-femininity test deemed that trainee priests and ministers were particularly feminine (Terman and Miles 1936, 153), and the test coded the desire to travel to "learn about various religions" as evidence of femininity (Terman and Miles 1936, 502). In *Psychological Factors in Marital Happiness* (Terman 1938, 230, 235), husbands and wives who had had a moderate amount of religious instruction had happier marriages both than those who had strict religious upbringings or no religious instruction at all. Wives' religious training had no impact on their later "orgasm adequacy," but more religious men were more likely to have "inadequate" wives (395–96).[23] Terman did not consider religiosity abnormal, but he did not consider it to be constitutive of mature masculinity either. In his letter to Yerkes in 1949 he described how "all theologies and dogmas seem to me about equally childish, including the Protestant, the Catholic, the Mormon and the Jewish."[24] Perhaps it was this ambivalence that led Terman to such actions as advising his student Harry Israel—who was not himself Jewish—to change his surname to "Harlow" to avoid anti-Semitic prejudice within the profession of psychology.[25]

If the value judgments of secular sciences are never fully exorcised, what can a psychology of religion do? Musing on this question at the end of his book, Graham Richards argued against psychologies which consider religions to be homogenous entities that could become an object of psychological knowledge by virtue of "the universality of spiritual yearning." I share his optimism that a psychology of religion would be more inclusive if it considered instead "a need for collective ritual . . . at the heart of even the most ascetic forms of religion" as being at the center of any psychology of religion (Richards 2011, 148). Post-Foucaultian psychologies are no exception. Will religion always remain a displaced ghost that haunts the social sciences, an available source of metaphors with which to shame those sciences with pretensions to value-free intellectuality? Perhaps an insistence on the centrality of collective ritual to such sciences provides a way to make sense of what is overlooked by Robinson's claim that "Kinsey was fully conscious of the demystifying motif in his work" (118). Perhaps such an insistence blasphemes against Foucault's confessional hypothesis, providing a counter-metaphor for the proliferation of secular and religious movements that channel the "will to truth" in the decades since Kinsey. Rather than a singular history of confessional practices, histories of the *scientia sexualis* might be written with an eye for their characteristic inclusions and exclusions, for the switches between those discourses of normalization and purification within the terms of the present, and those discourses that urge collective investment and which promise transcendence, deeper knowledge, or ideal futures. I am willing to take a leap of faith that such modern people as Kinsey and Freud who understand sex as organized by repression, and such postmodern people as post-Foucaultian psychologists who understand sex as organized by the production of discourses, might be similar to ancient people not because we have inherited their prescriptions directly, but because we have all been inclined to look back in time to religious forms of life that we understand poorly to invent metaphors and practices to orient our own collective lives through uncertain presents.

7

Frontier Living, by Figures Alone

Upon hearing about Kinsey's *Sexual Behavior in the Human Male*, the actress Mae West quipped that "when it comes to statistics, I'm afraid the only figure I employ is my own."[1] A very different sentiment was expressed with a similar metaphor two years earlier by Walter Shewart, the outgoing president of the American Statistical Association (hereafter, the ASA). Shewart's ominously titled speech, "The Advancing Statistical Front" (1946), was part of a tradition of speeches by outgoing presidents that stretched back into the nineteenth century. Therein, Shewart hitched the future of American statisticians to Vannevar Bush's ideal of an endlessly expanding frontier of scientific knowledge, and expressed an optimism typical of the speeches of outgoing ASA presidents in the years immediately following World War II—a period of growth for the membership of the ASA itself.[2] Shewart called on his colleagues to convince scientists, bureaucrats, and the public of the merits of the discipline: "If statistics can be made to serve each of us in a way that we individually recognize and appreciate, statisticians will no longer be referred to in some quarters as the showgirls of the market place because they live by their figures alone" (1946, 2).

Donna Haraway's histories of American biology position the postwar period as one in which information became the dominant mode of understanding life. More critical of this epistemological shift than Shewart, Haraway described an "informatics of domination" organized less by the Foucaultian politics of organisms and populations than by a cyborg politics of simulation in which "sexual reproduction" and "sex roles" ceased to be either material or ideological bedfellows.[3] For Haraway, the organization and funding of the life sciences for "sex, mind, and profit" are a turning point in the "translation of the world into a problem of coding" in the postwar period. Haraway describes psychobiologist Robert Yerkes as an exemplary biologist of the period before this translation. Robert Yerkes conducted research through the CRPS and the Committee on Scientific Aspects of Human Migration "for purposes of rational social management policy." In contrast, the era of the informatics of domination was shaped by leadership of the Division of Natural Sciences within the Rockefeller Foundation by Warren Weaver. Under Weaver's leadership, biology became a communication science whose objects were information flows rather than organisms. Haraway considered Weaver to be representative of the biologists of the later period.

Surprisingly, the statistical representation of human sexual behavior in Kinsey's studies is barely mentioned in Haraway's histories.[4] However, the Kinsey project flourished under Yerkes' chairmanship of the CRPS and became unstuck after 1950 when Weaver became head of the Rockefeller Foundation's Division of Natural Sciences. Few scientists were "caught up in the internecine battles of the Rockefeller Foundation, where the political currents were both strong and treacherous" (Jones 1997, 636), more than Kinsey. Moreover, this change in fate drew the frontierist imagination of the American Statistical Association to the sampling methods in Kinsey's work. American statisticians' historical agency is sometimes invisible in the Kinsey literature, wherein Kinsey's sampling is evaluated for bias and accuracy without regard for the lack of stable norms for "unbiased" sampling.[5] Much of the Kinsey literature remains in the state bemoaned by Ian Hacking in 1981 in which "there is nothing as anonymous as the bureaucracy of the statisticians" (1981, 24).

Because the debate between Kinsey and the statisticians remains contested and contestable, it seems worthwhile to narrate that interaction in some detail. Kinsey's study prompted the imagination of new normative ideals for a national survey of sex, but Kinsey also articulated resistance to the statisticians' ideas as to how a sample of the national population might be gathered, and how the resulting data might be interpreted. Today these matters are barely remembered as technical debates, internal to the social sciences. However, in the early 1950s they were not only cutting-edge cross-disciplinary debates but also matters of public concern and interest. I will first present a history of statistics in the United States that ends with postwar sampling theory, before examining Kinsey's interaction with the statisticians.

The History of American Statistics

Since Hacking bemoaned the anonymity of the statisticians, historians have traced how nineteenth-century statisticians invested governments, the public, and the scientific disciplines in counting practices, ontologizing "society" and "nature" as probabilistic objects.[6] Statistics informed eighteenth-century notions of enlightenment reason and nineteenth-century notions of *l'homme moyen*, both of which served as touchpoints for rationalist social policies. While historical research on statistics has been somewhat Eurocentric, many eighteenth-century Europeans noted that the American revolutionaries had placed a remarkable faith in democratic ideology by resolving to count people every ten years and to reapportion political representation to accord with the states' relative populations.[7] Several statistical societies formed in the United States during the nineteenth century, but the American Statistical Association—formed in Boston in 1839—was the only one that survived into the twentieth.[8]

As Shewart's use of the frontierist metaphor of the "advancing front" suggests, there are particularities to the relationship between statistical understandings and political government of the people of the United States. Nineteenth-century American statistics engaged in troubled ways with contested, racial definitions of American personhood. Native Americans were not counted on the census unless taxed, while Southern slave owners argued that their slaves should count as people for reapportionment purposes. Northern Yankees made similar arguments about counting new immigrants who often lived in conditions little better than slavery. Census figures were brought into antebellum slavery debates, as when some Southerners argued that the 1840 census suggested that more "insane negroes" lived in Northern than Southern states and used this supposed demographic difference to argue for the legitimacy of slavery. The miscalculations that led to this conclusion also led Northern ASA members to an increasing and enduring involvement in census design and execution.[9]

Shewart's metaphor was not the first time that American statistics and frontierist ideology intersected. While European statisticians of the early nineteenth century may have invented *l'homme moyen*, American statisticians of the late nineteenth century labored to locate the "average American" in space, as the center of gravity of the population moved ever westward.[10] By 1890, statisticians estimated that "the average American" was no longer moving, leading Frederick Jackson Turner to announce the end of the formative era of the frontier in United States history.[11] In the same period, the typical American *statistician* began to move westward, taking an educated intelligence of quantifying people with him. In the 1880s, the membership of the ASA spread far beyond New England, and in the 1890s statistics became a more routine part of the curriculum of American universities.[12] In *The Closing of the American Frontier*, Turner expressed a hope that the state universities where such statisticians worked might draw an expert class of scientists from the people of the Plains to avert the dominance of public life by experts in private universities. American public universities had "a duty in adjusting pioneer ideals to the new requirements of American democracy necessarily implicated in the stability of the states" (Turner [1920] 2011, 181). By 1900, the establishment of a permanent Census Bureau, replacing the occasional assembling of statisticians for the ten yearly counts, welded the profession of the American statistician to the quantification of the American population.[13]

To understand Kinsey's debates with the statisticians, it is necessary to understand the events that made sampling so pivotal to the authority of mid-twentieth-century statistics. Haraway described a shift in the mid-twentieth century away from Foucault's biopolitics—centered on national populations—to the more potent cyborg politics engaged with the invention of the artificial, displacement of the natural, and hybridization of the informatic and organic.[14] The emergence of *inferential* statistics in the early twentieth century deserves a pivotal position in the history of this shift. Following British statistician Ronald Fisher and his followers, American statistics in the 1920s expanded its scope from demography to include techniques emerging from agriculture and industry which attempted to stabilize doubt about the range of *populations* that might be inferred on the basis of observation of a small sample:

> From a limited experience, for example, of individuals of a species, or of the weather of a locality, we may obtain some idea of the infinite hypothetical population from which our sample is drawn, and so of the probable nature of future samples to which our conclusions are to be applied. (Fisher 1925, 43)

Because these research methods allowed a quantification of the probability that a particular sample in hand had been drawn from a particular population, they also allowed a quantification of doubt about the question of whether two samples had been drawn from the same population or not:

If a second sample belies this expectation we infer that it is, in the language of statistics, drawn from a different population. . . . Critical tests of this kind may be called tests of significance, and when such tests are available we may discover whether a second sample is or is not significantly different from the first. (43)

In other words, Fisher's inferential statistical techniques "inverted" the traditional relationship between populations and samples.[15] By this I mean that pre-Fisherian demographic statisticians ontologized populations as real things that they might sample through various methods. However, Fisher's methods led the sampled data in hand to appear more "real" than the populations from which those data had been drawn. Inferential statistics allows the statistician to quantify doubt about the inference that a sample in hand has been drawn from a population with particular characteristics. By so doing, inferential statistics professionalized and normalized the imagination of ranges of imaginary populations from real small samples. Crucially, Fisher's methods for managing such inferences and doubts assumed that population distributions occurred in bell-shaped curves.

Inferential statistics effected widespread change in the statistical profession and in the practices of American government. The American statistician Harold Hotelling (1927) announced Fisher's work as "revolutionary," and American statisticians debated the assumptions of Fisher's tests that natural variation occurs in Gaussian bell-shaped distributions. In 1927, outgoing president of the ASA Leonard Ayers voiced worries that the mathematical knowledge required to execute and interpret inferential statistics might create a "two-tier" system among American statisticians.[16] After the stock market crash of 1929, Fisher's statistical methods were put to work by government statisticians concerned with testing the effects of social programs. During the mid 1930s, ASA members were consistently enjoined by their outgoing presidents not simply to study national life but to socially engineer it.[17] The 1935 Social Security Act increased grants-in-aid to alleviate near-ubiquitous poverty, and statisticians increasingly asked questions about the effects of governmental interventions on the "standard of life." Because Fisher's inferential statistics leveraged small samples to answer questions about the likely effects of social policies on larger populations, they became a means of cheaply testing the quality of social policies aimed at increased poor Americans standards of life.[18] Echoing Hotelling's (1927) metaphor of a quarter century earlier, statistician W. J. Youden reflected in 1951 on the "Fisherian revolution" in the treatment of experimental data.[19]

The importance of sampling as a means of predicting behavior en masse was also publicly highlighted by the success of George Gallop in predicting the results of the 1936 presidential election. Gallop's smaller but better-sampled poll outpredicted the larger poll by the periodical *Literary Digest*, which had polled its predominantly middle-class and Republican readers.[20] Through such successes, Gallop expanded a market for opinion polls, claiming that they strengthened democracy by representing the public will to politicians more regularly than did the ballot box. However, as several historians have pointed out, the opinion poll rapidly became a tool by which politicians could predict and control the will of the people.[21] By the end of the 1930s, professional statisticians employed by the state were codifying the principles along which populations could be stratified and subdivided to manage the risk of "sampling error," learning the lessons of Gallop's 1936 victory.[22]

The Fisherian revolution even affected the procedures of the national census. For the 1940 census, additional questions about lifestyle, native language, and parental origin were asked to 5 percent samples of the population, abandoning the traditional norm of asking all respondents every possible census question.[23] Using the latest punch-card technology, the processing of these data was greatly accelerated. During the war there were several serious attempts to grapple with the ontological lessons of developments in sampling theory, deploying the army as a field of experimentation. In sum, the cyborg nature of American life was in the making through increasing academic, governmental, and corporate investment in sampling methods for rationalizing debate and doubt about the nature and characteristics of the national populations. Surprisingly, the largest national study of human sexual behavior took little account of these developments of the previous decades.

Kinsey's Resistance to Inferential Statistics

Because Fisher's influence on statistics was both anticipated and remembered as "revolutionary" by American statisticians, it is perhaps worth remembering Thomas Kuhn's claim that scientific revolutions are not clean historical breaks: "there are always some men who cling to one or other of the older views, and they are simply read out of the profession" (1970, 19). Raymond Pearl was ASA president when he visited the Indiana University campus and befriended Kinsey in 1939. Pearl had long been in the business of predicting the characteristics of future populations, but his biometric approach to demography did so on the basis of presumed laws of natural population growth rather than through inferences made from small samples to hypothetical populations. His was a neo-Malthusian biological science more firmly wedded to biopower than to cyborg artificiality. Pearl advised Kinsey that he need pay little attention to developments in sampling theory, describing sampling as a poor substitute for the collection of large samples. He also referred Kinsey to a natural scientist, the astronomer Frank Edmundson, for future statistical advice. This relationship between Kinsey and Pearl did not have the chance to develop; Pearl died suddenly and unexpected of a heart attack in 1940.

In spite of Pearl's premature death, his brief friendship with Kinsey had a lasting influence on the methodological debates that would surround the sex surveys that Kinsey was beginning to undertake. Pearl's junior colleague Lowell Reed was invited as a statistical expect by George Corner and Robert Yerkes of the CRPS to accompany them on their visit to the IU campus in December 1942. After his visit to Bloomington, Reed expressed very different views to Kinsey and to Yerkes as to the soundness of Kinsey's techniques. Three days before Christmas 1942, Reed wrote to Yerkes to recommend that Kinsey develop a scoring system for his items and that his code should be placed in a safety deposit box.[24] Reed also recommended that Kinsey hire another statistician. Perhaps anticipating such methods as "proving the answer," Reed pointed out that "one or two technics that he is employing at the moment would be subject to criticism with regard to his treatment of unanswered questions, or of questions with which the individual had had no experience." Reed continued that "it might be desirable to make available to him the funds with which he could hire another assistant."[25] Six days later, Reed wrote to Kinsey with a more reassuring tone: "I think you are doing a fine job. I made one or two suggestions with regard to statistical treatment while I was there, but I think the most important thing is a more thorough review of your grading scales."[26] In August 1943, Kinsey wrote to Reed that he needed advice on hiring statistical help: "we don't need a statistician who would not be willing to spend the major portion of his or her time in the running of machines, making charts, etc."[27] Reed replied that Kinsey did need only a "working clerical statistician" who might know some statistical procedures.[28] Clyde Martin was retired from active interviewing and came to do this clerical work, transforming Kueffel and Esser General Data Sheets into punchcards into the output of the Hollerith machine into tables and into accumulative incidence curves.[29]

Haraway has described the CRPS under Yerkes as "set up to study human variability for purposes of rational social management policy" (1991,

49). However, in the 1940s, Yerkes and Kinsey had different views on the importance of sampling "normal" people for Kinsey's study, and the difference was evident in the views each man expressed about sampling in their private letters. Voicing reluctance ("lest I disturb your normal functioning!"), Yerkes suggested that Kinsey confine his study in the first instance to "presumptively typical and normal male and female whites of the U.S.A. culture and between the ages of 12 (?) and 70 (?) years," deferring inquiry into "human variants" until a reliable picture of the norm had been built up.[30] Kinsey replied to reassure Yerkes that he was primarily interested in norms and averages rather than exceptions. However, the natural phenomena he was studying required a suspension of the increasingly conventional Fisherian wisdom that variation was "normally" distributed. "There is no modal portion of a population in curves which are shaped as most of ours. Out of 18 people who sat around our table in New York, there are not more than 2 or 3 who would fit any one person's notion of what is usual and modal in human sex behavior." In justifying his developing intelligence of sexual behavior, Kinsey positioned himself "as a taxonomist and a student of variation, securing the picture of the population as a whole" and as a trustworthy observer, someone "with some knowledge, for I have the histories of most of that group."[31] Contrary to Haraway's characterization of Yerkes' management of the CRPS as singularly orientated towards rational management, he trusted in Kinsey's emerging project with its novel approach to sampling, typicality, and taxonomic understanding.

Kinsey continued to invest in Pearl's notion that random sampling was a poor substitute for large samples.[32] He cited Pearl as support for the notion that sexual behavior was *not* normally distributed, as Fisherians' tests would assume (*SBHM*, 89), noting that it was "customary in statistics" to measure the accuracy of mean values using their standard deviations as measures of their likely variation (88).[33] The taxonomic method did not aim to measure mean values and to quantify doubt about the range of that population mean with respect to its variance. Rather "the student with practical experience in taxonomy" would know that the addition of a few cases to small samples can "alter the original calculations to an extent which would never have been anticipated through an examination of the standard deviations of the means" (89). Empiricism is at its most shaky when making inductions from observations of "what happened" to claims about "what happens" in a general sense.[34] Kinsey favored the *stratified sampling* of predetermined subgroups of the population over the *representative sampling* of samples drawn at random from the population as a whole. This decision was also informed by a taxono-mist's experience with variability; representative sampling led to unreliably small samples.[35]

Kinsey's sampling strategy was tied to his sex survey and as a result he defended it at length in *SBHM*. To manage the disparity between a student-heavy sample and the United States population, Kinsey developed "US cor-rections" that weighted his data according to the 1940 Census breakdown of the US population along lines of sex, race, marital status, age, number of years of schooling, rural-urban background, religious group, religious adher-ence, and geographical origin (*SBHM*, 104–9). Kinsey acknowledged that his corrections involved some guesswork and extrapolation, but argued that he had done a *better* job than other social scientists: "it is unfortunate that stu-dents in psychology and the social sciences regularly publish raw data with-out corrections for the Census distributions of their populations" (109).

Kinsey was also aware of a weakness that haunts the sampling strategy of sex surveys, in which participants are volunteers. Volunteers must consent to talk about their sex lives to researchers. By depending upon the free choices of participants, sex research allows a reasonable doubt that volunteers might be particular, unrepresentative kinds of people. Sex researchers such as Davis, Hamilton, and Kinsey had brokered the intelligence, education, and rationality of their participants to avert criticisms of their participants' abnormality. However, within the ontology of representative sampling, the participants in sex surveys do not need to be abnormal, but merely atypical, for that science to be compromised. To control doubt about the atypicality of his sample, Kinsey compared "partial samples" with his "100% groups" in which all members of particular groups had been sampled. He emphasized that the differences in rates of particular experiences between these groups were minor when compared with the deviation of the overall sample from cultural expectations (*SBHM*, 98–99).[36] In contrast, the 100 percent groups tended to show slightly lower weekly rates of sexual behaviors, and Kinsey struggled to alight on a single explanation for the disparity. People may have been moved by cognitive dissonance to talk more freely about sex as a result of participating in sex research, but people who were more varied in their sex lives were more likely to consent to talk to sex researchers in the first place.[37] Many people were enrolled into Kinsey's study and became more curious about sex surveys as a consequence. But some chose not to make this initial investment in the endeavor. Finally, Kinsey also conducted repeated samplings of his own data to establish, inductively, the minimum number of people of a particular social grouping that it was necessary to sample to achieve a reliable estimate of the population's characteristics (*SBHM*, 82–92, 736–65).

Gentlemen's Disagreement?

Kinsey's sampling strategies occasioned praise and blame. Social science reviews of *SBHM* noted disparities between Kinsey's sample and the population of US White males.[38] However, while psychologists Goldstein and Pastore critiqued Kinsey's volunteer sample, opinion pollsters Hymen and Sheatsley attributed Kinsey's success in achieving rapport to "diligent perseverance and long-established relations in the community" (1948, 185). Goldstein and Pastore (352) also described the interview methods as likely to mislead people, particularly those to whom Kinsey had attributed feeblemindedness. Hyman and Sheatsley highlighted similar concerns but concluded that "in a work of this monumental scope, to harp on such small matters would be captious" (1948, 195). Writing in *The American Journal of Psychiatry*, sociologist A. H. Hobbs and his colleague R. D. Lambert used the sampling strategy to impugn Kinsey's intentions: "disproportions within the sample are approximations because of the unconscious or deliberate failure of the authors to include the actual numbers of cases involved in the various categories" (1948, 759). Similarly, Terman (1948, 447) accepted Kinsey's explanation as to why the ideal of random sampling could not be achieved in a sex survey, but critiqued Kinsey for failing to communicate the size of various social groupings effectively. In sum, Kinsey's sex survey was obviously broader in scope than previous studies, and it consequently affected the imagination of social scientists as to what a *scientia sexualis* might be. Through their discussion of what Kinsey had and had not achieved, social scientists articulated a new ideal of a representative national survey of the nation's sexual behavior. Although Kinsey's work prompted the imagination of this ideal, the particulars of his sampling strategy also meant that his survey could be seen as falling short of it.

While Terman's review was not distinctive in calling attention to the ways that Kinsey's sampling might be less than ideal, he rather uniquely stepped aside to allow another gentleman to voice the criticism: "the reviewer asked Dr. Quinn McNemar to check over the statistical procedures used in this experiment." We last left Terman's student Quinn McNemar in chapter 5, beating out E. Lowell Kelly to the job of analyzing the data for *Sex and Person-*

ality, but his subsequent career is part of the story of the Fisherian revolution in psychology. McNemar completed his Ph.D. under Terman in 1932 and became a faculty member at Stanford thereafter in Psychology, Statistics and Education. Harold Hotelling introduced McNemar to Fisher personally in 1933, and McNemar learned the method of analysis of variance for analyzing the results of experiments from Ronald Fisher directly. In the 1930s, McNemar had the privilege of studying with many of the leading statisticians of his day, including the Princeton mathematical statistician Samuel Wilks. In the early 1940s, while Kinsey was gathering sex histories, McNemar was living in New York and running the grants-in-aid program of the Social Science Research Council while his wife Olga was completing her Ph.D. in psychology at Columbia University.[39] Through his personal teaching and through his textbooks, McNemar solidified the place of Fisherian statistical techniques, particularly the analysis of variance, in stabilizing the interpreting of experimental data in psychology from the 1940s onward.[40]

Quinn McNemar wrote two lengthy reviews on social science methodology in the early 1940s that Kinsey cited in support of his own taxonomic method of sampling.[41] As noted in chapter 3, the McNemars had met Kinsey at the 1946 meetings of the American Psychological Association, and carried their good impressions of Kinsey back to Terman at Stanford.[42] However, McNemar's review-within-a-review critiqued Kinsey's method of sampling his own data to estimate ideal sample sizes. McNemar concluded that "incognizance of four elementary statistical procedures renders worthless this elaborate effort to determine how large N should be for a subgroup" (Terman 1948, 450–51). With Kinsey's science so publicly undermined by the very statistician whom he had cited for support of his sampling strategy, Terman's review continued with the four particular criticisms that I have used to organize earlier chapters of this volume.

In May 1948, Terman wrote to Yerkes that he had "about finished my review of Kinsey" and was "trying very hard to take out of the comments anything that might wound his amour propre too much."[43] In the month after the review was published Yerkes replied that the review was "the first thorough-going and obviously careful and competent review that has come to my attention." Yerkes signaled that he did not agree with all of Terman's arguments, but described the review as "gentlemanly and considerate . . . even if a summary of defects and shortcomings." Yerkes even joked that Kinsey, the NRC, or the Rockefeller Foundation might do well do send Terman a one-thousand-dollar note in recognition of his contribution.[44] Terman replied quickly to express his relief that Yerkes appreciated the review which he had "worked very hard to make . . . accurate, objective, and as you put it, gentlemanly."[45] Terman noted that his opinion had been reinforced by Goldstein and Pastore's review, which he drew to Yerkes' attention.

Yerkes did not reply to Terman again for another two months, but neither psychologist neglected the review in the interim. Yerkes had recommended that Terman forward copies of his review to all members of the CRPS, and Yerkes himself sent a copy of the review to the Committee's new chairman, George Corner.[46] Kinsey wrote to Yerkes to assess his opinion of Terman's review, and Yerkes replied with lengthy and considered comments describing Terman as honest, competent, and fair-minded, and the review as objective. Yerkes advised Kinsey not to reply to Terman in print but to contact him in private. He concluded: "My chief hope and object in writing this letter is to help to bring Kinsey and Terman together as able and honest fellow biologists who should work as friends in spite of disagreements."[47] A week later, Yerkes broke his silence with Terman and wrote to him to garner his sympathy for Kinsey's efforts.[48] In reply, Terman put the responsibility for the outcome of the controversy firmly back on Kinsey's shoulders. "How much good the review does depends entirely on Kinsey's willingness and ability to profit from criticisms. On that I am not too optimistic."[49]

Yerkes clearly hoped that these two would behave as gentlemen and reestablish consensus over the vexed category of sexual science. His optimism was not to be rewarded. George Corner had succeeded Yerkes as chair of the CRPS in 1947. After reading Terman's review, he discussed it with Kinsey in person. Unlike Yerkes, Corner did not consider the review gentlemanly, found "it difficult to understand how an experienced psychologist could commit the ineptitude of frankly writing comments solely on the shortcomings and inadequacies of a fellow scientist's work. With this approach he could not possibly avoid the impression of ungenerosity, in spite of his disclaimers." Corner considered that Terman had been unfair to Kinsey at several points, including his and McNemar's analysis of the sample size problem. In other words, Corner was already thinking about how statisticians would judge Kinsey's work, and the review seemed overly critical to Corner because "Lowell Reed and Kinsey's other statistical advisors seemed to have been satisfied with his methods."[50]

In other words, the Terman review prompted doubt at the highest level within the CRPS as to what the standards of "good science" might be in the domain of sex research, and the CRPS had limited purchase on the matters at variance between these two. Was Kinsey or Terman the more valuable sex researcher? How could standards of "good science" that could demarcate the work the committee supported be reestablished? As well ask a horse trader how he knows when to close a bargain! Instead of appealing to such standards, the gentlemen of the CRPS appealed to the gentlemen of the American Statistical Association to arbitrate this debate.

The Statistical Front Advances to Bloomington

On November 8, 1948, Terman was contacted by W. Allen Wallis, a University of Chicago sociologist who had participated in Terman's Monday night seminars some years earlier. Wallis was organizing a symposium at the meetings of the ASA in December on problems in the Kinsey report. Wallis's praise of Terman's review was high: "I got the impression that you had said everything important, and that there is little point to the Cleveland session on Kinsey."[51] The executive secretary of the ASA, Merrill Flood, also invited Kinsey to speak at the Cleveland event.[52] Independently, Terman and Kinsey declined these invitations. While Terman rarely accepted invitations to travel to speak throughout his career, Kinsey learned that Lowell Reed would speak after the symposium and wrote to Flood that "that will take care of our side of the program."[53]

However, at the actual symposium, Reed did not take Kinsey's side, but openly criticized Kinsey's sampling. Reed later wrote to Kinsey to suggest how events might have turned out differently: "If the type of statistical guidance had been provided that I had in mind, I feel sure that you would now be free of some of the criticism that is now being justly leveled at the work."[54] In Kinsey's view, Wallis had not behaved like a mature gentleman in drawing together this symposium. He described Wallis to Reed as someone who was "young and cocky, and pushes himself forward in airing his opinions." He continued: "Frankly, I think the Statistical Association might have done better in arranging for a sound discussion of our problems."[55] Wallis may have been making hay while the sun shone on Kinsey. Reed chose to reveal his doubts about Kinsey's methods in this public forum before revealing them to Kinsey. Kinsey reacted defensively to criticism of aspects of science that he was ill-equipped to defend. While all parties in this matter appealed to notions of fair conduct, none were above somewhat ungentlemanly conduct.

After the ASA symposium, Terman put Kinsey under pressure with increasing boldness. He spent two days contributing criticism to a critical review of Kinsey by the young sociologist Paul Wallin, to be published later in April 1949.[56] Terman wrote to psychologist Albert Ellis to critique *his* review of Kinsey as insufficiently critical and for "lumping Kinsey's critics together and lambasting them without making a distinction."[57] Ellis replied that he had, on Kinsey's advice, toned down his criticisms, because he thought it more important to defend sex research as a whole against "anti-scientific forces" than to score points against him.[58] This account led Terman to reply to Ellis that Kinsey had "played you for a sucker."[59] Criminologist Norman Fenton wrote in appreciation of Terman's review, and to tell Terman that Kinsey would soon begin working in California prisons. Fenton had earlier advised Terman on how the M-F test might be used as a means for identifying suspected homosexual boys in the juvenile justice system.[60] Terman advised Fenton to investigate Kinsey's methods, either by listening to an interview from behind a screen or even by being interviewed himself.[61]

In public, Terman's review was also positioned as original and authoritative by statisticians. Wallis published his review of *Sexual Behavior in the Human Male* in the *Journal of the American Statistical Association* in December 1949. In that review, he restated the now familiar doubts about Kinsey's sampling and his interview methodology, praised Terman's review unequivocally, and made public the original 1942 letter which expressed Lowell Reed's doubt about Kinsey's methods. Kinsey's taxonomic method had now visibly fallen short of a new statistical ideal of a national sample of sexual behavior, and the statistical shortcomings of his work had an available historical explanation in Reed's letter. The scientist in whom the CRPS had singularly invested human trust and financial resources throughout the 1940s was on the one hand wildly popular with the public and on the other an object of criticism by statisticians who claimed new particular expertise in representing that public. Moreover, these statisticians were bestowing great respect on the views of Lewis Terman, the friend of the CRPS's longstanding chairman, who had previously been granted trust and money to research human sexual behavior. *SBHM* had opened up the nation's sexual behavior as a new frontier for the advancing statistical front.

Early in 1950, the conservative Warren Weaver joined the Rockefeller Foundation as head of the Natural Sciences Division and brought increased scrutiny to bear on the practices of Alan Gregg, the head of the Medical Division. Gregg had publicly supported Kinsey, had written the introduction to *SBHM*, and had supported the funding of the Kinsey project via the CRPS. In May 1950, George Corner asked ASA President Samuel Wilks to evaluate the Kinsey research. Prior to the publication of *SBHM*, some members of the American Statistical Association had already begun to imagine a new goal for statisticians in correcting public misunderstandings about statistics.[62] By

1948, a Committee on Statistical Standards and Organization of the ASA had a chairperson and was in the process of selecting its members.[63] The public's growing interest in statistical representations of sexual variation seemed like a territory to which statisticians might orient themselves. Prior to taking on the job of auditing the Kinsey project, Samuel Wilks phoned his friend Warren Weaver, who urged him to proceed.[64]

Kinsey and Wardell Pomeroy first met with Wilks in July 1950 to agree the terms by which a committee of ASA members would audit the Kinsey team's work. At this meeting, Kinsey and Pomeroy effected a very successful piece of ontological gerrymandering. In a detailed letter that recorded their agreement with Wilks, Kinsey first recorded that the auditing committee would "consist of three or more persons, all of whom should be at Bloomington for a week or more to make a first-hand examination of our methods." Such people would be primarily *statisticians*. "The choice of a statistician because he is basically a psychologist, a psychiatrist, or something else is not in point." Second, the members of the committee would be individuals who had *not* previously published criticisms of Kinsey's work. The statisticians would spend time getting acquainted with the overall approach of the research and would contribute their own sexual histories in the process. Such initiation was presented as essential. "There is no other way in which we have done our interviewing, and since our interviewing techniques have been questioned by our critics, we think it highly desirable that the committee have a first-hand knowledge of the interviewing." Third, the review would be oriented around the published criticisms of Kinsey's work. "Since there are only four statistical reviews which are basically involved . . . it would not involve too much work for the committee to make such a review." Kinsey insisted that *SBHM* would be assessed with respect to the descriptive norm of existing surveys of sexual behavior and not simply with idealized methods that had yet to be actualized in practice; the committee would also "compare methods in our research with methods in other published research in similar fields." Finally, the Kinsey group would have the right to respond, in the presence of the NRC Committee and in Bloomington.[65]

These terms remained contested as the committee of William Cochran, Frederick Mosteller, and John Tukey was convened in September and visited Bloomington first in October 1950 to carry out its work. To be sure, each statistician gave his sexual history as agreed. But when Cochran later pressed Kinsey to explain his interview method in greater detail,[66] Kinsey insisted that he was too busy with other matters and that he wanted to remain focused on studying sex rather than on methodology.[67] After this visit, the three statisticians met repeatedly in Chicago, Princeton, Cambridge, and Baltimore, and they visited IU again on February 23 and 24, 1952 to present and debate their assessment of *SBHM*. George Corner, William Allen, and Carl R. Moore were also present to represent the National Research Council. Statisticians' sampling ideals remained relevant to this discussion; John Turkey and Kinsey exchanged particularly frank words at this meeting over the relative merits of small systematic random samples versus large opportunistic ones.[68]

The book-length review (Cochran et al. 1954) that these three statisticians reported was more than simply an audit on the Kinsey work; it was a strong statement of the ideals of sampling theory—a movement of the statistical front across the virgin scientific territory that Kinsey's sex survey had led so many scientists to imagine as a possibility. Long sections of the book detail the assumptions of sampling theory with no reference to Kinsey or sex research whatsoever (233–36, 249–54, 309–31). Kinsey's work was now positioned as outmoded and in need of new disciplinary colonization lest the nation be misrepresented and misled by dubious sex surveys. While Kinsey had described taxonomy as "the oldest of the biologic sciences," the statisticians critiqued Kinsey's taxonomy for evincing "no detectable semblance of probability sampling ideas" (12). Fisher's inferential statistics had suggested that his methods for analyzing variation cut across species boundaries when he described how they " have nothing to do with the political unity of the group, and are not confined to populations of men or of social insects" (Fisher 1925, 2). Cochran and his colleagues similarly detailed how sampling theory ought to modernize and render redundant once acceptable taxonomic practices of making sense out of biological variation:

> In the early years of the present century, it was not uncommon to measure the claws and carapaces of 1000 crabs or to count the number of veins in each of 1000 leaves, and to attach to the results the "probable error" which would have been appropriate had the 1000 crabs or the 1000 leaves been drawn at random from the population of interest. If the population of interest were all crabs in a wide-spread species, it would be obviously almost impossible to take a simple random sample. But this does not bar us from honestly assessing the likely range of fluctuation of the result. Much effort has been applied in recent years, particularly in sampling human populations to the development of sampling plans which, *simultaneously* (i) are economically feasible, (ii) give reasonably precise results, and (iii) show within themselves an honest measure of fluctuation of their results. Any excuse for the practice of treating

non-random samples as random ones is now entirely tenuous. (Cochran et al. 1954, 14).

Future sex researchers who might wish to position themselves as frontiersmen, as Kinsey had done—and who might aim to assert that "we have done as well as anyone knows how to do"—were directed towards improving sampling methods above all else (1954, 264). The future imagination of the genre of the sex survey was being shifted away from a taxonomic study of organisms toward a cyborg statistical science based on the rational imagination of a range of possible national populations.

However, the statisticians' published final conclusions on *SBHM* were milder than many of the other social science criticisms, including Terman's. *SBHM* was described as the best of available surveys of sexual behavior, and Terman's *Psychological Factors in Marital Happiness* as its closest contender. Without an explicit account of the interview to critique, the statisticians behaved as Kinsey would have hoped; they reported—in largely positive terms—on their own experiences of having their sexual histories taken. They also acknowledged that sex was a topic that was ill-suited to empiricism; until new methods were found, sex would continue to be studied via self-report, with all of its attendant dynamics of faith and doubt (1954, 3). However, Cochran, Mosteller, and Tukey also recognized that Kinsey's methods for spreading trust in sex research through social networks had merit. It was an easier way to recruit participants than random sampling, and random sampling might increase the financial cost of securing an interview by a factor of between five and fifty (1954, 55 and 263).

The Cochran, Mosteller, and Tukey review was certainly the most exhaustive discussion of the sampling questions prompted by Kinsey's achievements. The ASA further established itself as the ontological background by which such controversial debates among social scientists would be thrashed out with a symposium on the book published in the *Journal of the American Statistical Association* in 1955, edited by Wallis (Kinsey et al. 1955). In his introduction, Wallis described Cochran, Mostellor, and Tukey's book as "a credit not only to its authors but to the Association that sponsored it and the officers who selected the authors" (1955, 827). In response to the summary of the books' conclusions published in this symposium, Kinsey, Pomeroy, and Martin defended the rationality of their ongoing sampling practices, describing Cochran, Mosteller, and Tukey's ideals as insensitive to the practical difficulties of securing rapport in sex research (811–14). Terman's contribution (824–25) consisted of a short list of editorial criticisms, printed verbatim from a private letter, that appear to have not been intended for publication at all. Cochran, Mosteller, and Tukey's contribution had the last word, concluding that "although we are not in complete agreement with all the points made" by the other reviewers, that "our remaining differences hinge mostly on questions of judgment, emphasis, or frames or reference" (829). Order had been returned to the sexual sciences, reconstituted on statistical grounds.

The advancing of the statistical front over Bloomington was also a moment to advertise the authority of the ASA to a wider public. In June 1952, someone, somewhere, leaked to the press that a statistical audit of Kinsey's work was taking place, leading Kinsey and Corner to rush a release that summarized the statisticians' findings.[69] This event prompted one of those rare moments in which the rationality of different sampling strategies became front-page news in local newspapers. The Kinsey archives in Bloomington contain 117 newspaper clippings that picked up the story, largely published on June 25, whose headlines ranged from "Kinsey Failed" in the *Burlington North Carolina Times-News* to "Kinsey Report Wins Praise: Statisticians, however, say it's not fully accurate" in the *Baton Rouge Louisiana State Times*. Most articles noted that sex research was rendered less than ideal because of the difficulties of sampling, and by the need to trust what people said they did in private. Only a few went as far as the *New York Post* in tutoring its readers on different ways of ontologizing the nation through sampling:

> The major disagreement between the statisticians and Kinsey is over polling procedures. Generally the statisticians recommend a method called "random sampling," in which the choice of subjects to be interviewed, depends only on chance. Kinsey, on the other hand, has favored a kind of "stratified sampling" or selection of subjects according to a prearranged plan.[70]

The *Ann Arbor Michigan News* was typical in citing the report's opinion that "for some of the statistical problems Kinsey encountered there just aren't any known solutions." The statistical front both advanced over the embodied practices of sex research and led to new imaginations of better sampled research that belonged to the future. Kinsey's work may have led people to invest in the truth of his science, but he had been exposed as a less-than-ideal scientist, and certainly not one who could address all of the criticisms of the statistical pioneers. However, the judgment of Kinsey by the statisticians of his time was more measured than some subsequent judgments that his studies were characterized by "bad sampling." The space between what Kinsey had done and the possibility of what *could* be done was something that might be visualized by statisticians on behalf of the public—statisticians who might represent the totality of the nation with confidence. The statistical front had advanced and order had been restored. Only matters of opinion remained: reasonable things for gentlemen to disagree about; minute details that might become of interest to historians in later decades.

8

Normalization Now

I started this book with the Foucaultian idea of genealogy, obliging myself to repay readers' attention with some unsettling of apparent epistemological and ethical certainties of the present. This short and final chapter is my attempt to settle any accrued debts. Herein, I return to the psychology of the past twenty years, since I was initiated into the group practices of Stanford psychologists in Terman's department. Throughout, I have described two kinds of normativity, which value averages and ideals, and which have shaped the discourse of human differences in the human sciences. The former Queteletian understanding has adhered more to sexuality, and Foucault's discussion of "disciplinary power" in the psy- disciplines describes the form of power/knowledge in which it is invested. Galtonian normativity has adhered more to intelligence, and it is oriented toward the visualization of ideal forms of life; it is a form of power/knowledge in which those visualized are not always equivalently entrapped. In particular it allows the possibility of escape from normalization when people are successfully imagined to embody rational scientific thinking. My thesis has been that the history of psychology has been shaped by the intersecting operations of these two very different logics of normalization much more than has been acknowledged. The history of power/knowledge has been one of envisioning hopes and ideals for collective futures, conservative efforts to maintain the status quo from the destabilizing influence of atypical people and events, and switches between one frame of reference and the other. The Kinsey–Terman debate can be understood from this vantage point on history but remains a puzzle without it.

If this argument is right, then the normalization of sexuality and intelligence ought to continue to intersect in material ways in our own times.

One of the most influential gender theorists of the 1990s, philosopher Judith Butler, described ontologies which presume "basic" or "original" forms of heterosexual genders which had excluded queer genders at foundational moments and later came to serve as norms for "gender," which lesbians and gay men's genders were said to copy—often poorly (Butler 1990, 1993). Butler's schema resonates not only with the gendering of the instincts in psychoanalysis that she describes but also with the gendering of items as "masculine" and "feminine" on the M-F scale described here.[1] In both cases, entities such as scale items and discourses about pre-Oedipal instincts became endowed with the power to name individuals' subjectivities as "masculine" or as "feminine" within a heteronormative space that allows only heterosexual genders to be originary. My reading of Terman and Miles's *Sex and Personality* and the associated elements of the GSG series resonate with Butler's reading of the ways that Freud's work makes queerness visible so that it can serve as a pivot point for modern power. Other historians have documented how later measures of masculinity-femininity drew on the logic of *Sex and Personality*, associated common responses of women and men with "masculinity" and "femininity," and used these normalizing terms to support the diagnosis of homosexuality as a mental illness, often within psychiatric settings in which psychoanalytic influence was dominant.[2] In many ways, the 1973 psychometric critique of such measures by feminist psychologist Anne Constantinople anticipated Butler's critique of the homophobic impulse at the heart of all attempts to ontologize "masculinity" and "femininity" in singular terms.[3] Such measures of "masculinity-femininity" remain in use among sexologists in mental health domains where children's expression of gender can be a basis for a psychiatric diagnosis.[4] This is not to say that the current pathologization of children's gender in psychiatry is *only* a consequence of homophobia. Terman's preoccupation with the adjustment of X/Renwick supports Karl Bryant's specific conclusions that the emergence of "Gender Identity Disorder in Childhood" depended on persistent "cisgenderism" that cannot be reduced to a secondary reaction against, or subtype of, homophobia.[5]

Terman's hopes for both X/Renwick and Henry Cowell suggested how the sustained Queteletian normalization of children's gender and sexuality is sometimes suspended by hopes for a Galtonian sublimation of homosexuality into intellectual works. How might this genealogy help to make sense of the "queer theory" of the past quarter century, which has attempted sustained analysis of lesbian, gay, and queer cultures that are not affected by heteronormative reference points? David Halperin (2005) has argued that "queer theory" became normalized when the term "queer" became secondary to "theory," so that "queer theory" remained within a disciplinary space rather than traversing across disciplines. I tend to agree that, to keep queer theory subversive, we must resist speaking in ways that are intelligible only to those who are initated into particular traditions in the humanities. For example, recent queer theory has critiqued homophobic ideologies that can limit freedom in the present by getting people to invest their hopes in fantasies about the future that are "figured" by children.[6] I remain curious about the subversive potential of such moves, but want to imagine new frontiers in which queer theory also engages with psychological work such as that conducted by David Perry and his colleagues. Perry's measures of gender identity could be easily written off as normalization-as-usual in the psy- disciplines. But his measures aim to distinguish the difference between the extent to which children feel different from the Queteletian norm for their gender group and the extent of the pressure that they feel to conform to prescriptive ideals of gendered conduct.[7] Such engagement across disciplines seems unlikely at present, but not because there is not nuanced work on normalization, gender, and children going on in multiple intellectual traditions. Rather, statistical and theoretical incommensurability obscure the queer possibilities in distinguishing something like Perry's work from that of sexologists who continue to defend the measurement of one-dimensional "masculinity-femininity" scales initiated by Terman. I hope that this history of Terman and Kinsey's incompatible sciences leads to de-investement in the notion that describing psychological work as "disciplinary power" is an adequately descriptive or critical engagement with psychology, and to more sustained scrutiny of the very different things that members of my group do in our experiments, surveys, case histories, statistical analyses, and data visualization practices.

Much as the contemporary normalization of sexuality incorporates the politics of intellectuality, it is difficult to look at psychology's most widely known discourses about the intellect without spotting familiar sexual politics at work. In our own time, psychological studies on the relationship between IQ scores and "spermin" would seem an embarrassing vestige of that earlier Hallian phase in the field's adolescence. However, recent literature on testosterone and IQ shows a recapitulation of Hall's hopes and fears about the essence of maleness. In some recent articles—in which race differences among men are not in focus—androgens have been found to be positively associated with intelligence among men but not among women.[8] Androgens are imagined as associated with the distinctly "male" intellectual traits thought to characterize autistic children. However, these days it is the finger-lengths of these children rather than the kinky pubic hair of gifted children that attracts the anthropometric gaze.[9] In the much more marginal psychology of racist J. Philippe Rushton, testosterone is purported to have an opposite effect as a "master switch" that can also contribute to aggression, selfishness, criminality, and sexual conduct (2000, 81). Rushton has argued that Black, White, and Asian people—imagined as three distinct races—vary in their performance on IQ tests for biological reasons (Rushton and Jensen 2005). Intelligence, sexuality, and crime are linked for Rushton because of inherited biological tendencies to reproduce often or to invest heavily in children's nurture. Rushton writes: "With higher testosterone levels, Blacks are more likely to put time and energy into having offspring. On the other hand, Asians and Whites with lower testosterone levels put more time and energy into caring for a few offspring and making long term plans" (2000, 81). I only mention Rushton here to make the point that when psychologists study testosterone *without* consideration of race, it often seems a good thing. However, when race differences among men loom large, this substance appears more like the embodiment of all that is worrying about men. The field of psychological science has perhaps retained embarrassing fantasies about race, desire, and the embodiment of maleness from its Hallian adolescence after all.

Rushton's theories about reproductive strategies also demonstrate the persistence of Terman's idea that high intelligence is a necessary prerequisite for the enactment of sexual morality into the twenty-first century. This hypothesis was most baldly stated in recent decades in *The Bell Curve* by psychologists Herrnstein and Murray. Consider how Terman's hypotheses about the lower IQs of prostitutes returned in Herrnstein and Murray's view that unmarried women with children have lower IQs: "The less intelligent a woman is, the more likely that she does not think ahead from sex to procreation, does not remember to use birth control, does not carefully consider when and under what circumstances she should have a child" (Herrnstein and Murray 1994, 179). Women with lower IQ end up on welfare because "the smarter the woman is, the more likely she will be able to find a job, the more likely she will be able to line up other sources of support (from parents or the father of the child), and the more farsighted she is likely to be about the dangers of going on welfare" (193–94). *The Bell Curve*'s authors argued that the "cognitive underclass" in the United States bore the brunt of a social "revolution" that of the late 1960s was characterized by a rise in divorce rates and births out of wedlock. The discourses described in chapter 2 are still with us.

Sander Gilman's (1996) argument that *The Bell Curve* was a book about morality, not intelligence, is the most insightful among the many criticisms of the book. Gilman recognized that the ontologization of groups as "naturally smart" can be as harmful as the ontologization of groups as condemned by genetics to a life of stupidity. Gilman describes how theories of high intelligence among Jews tend to mark that intelligence as derivative, as cunning rather than originating, and as witty rather than profound. Indeed, in 1994, the year that *The Bell Curve* was published, *Quiz Show* was nominated for best film of the year. *Quiz Show* offered a vision of smart people as devious, cunning, and discernibly Jewish, a reminder that, as Gilman notes, Jewish men are often made to embody the projection of all that is "queer" about the intellect. Like Butler's critique of the ontologization of heterosexual genders as originary, Gilman notes that the violence of "othering" smart Jews is not about the specific form of difference attributed to groups but about the ways that ontologies set up particular groups as the explanandum of a range of very different, and even contradictory, attributions of difference.

Smart Jews were not the only intellectuals othered and undone by *The Bell Curve*. Herrnstein and Murray's argument also targeted middle-class "liberals" caught up in the sexual revolution, influenced perhaps by Kinsey in addition to the poorer Americans whose IQ scores were displayed in multiple tables and graphs. The book argued that the "cognitive underclass" had not the intellectual capacity to deal with the new freedoms opened up by the sexual revolution. This obvious, longstanding association of differential intelligence with sexual control and conformity to sexual norms bears comment. And yet—perhaps attesting to the inability to think the relationship between normativity of sexuality and the intellect—there has been little by way of queer critique of *The Bell Curve* among the volumes of reaction to this work.[10] *The Bell Curve's* authors may have known better than many of the book's critics that middle-class Americans did not really fear a nation of people with low IQ as much as they feared a decline in sexual morality. It seems to me to be an obvious point that the racism and classism that animate this book had but little concern that poor, Black, and Latino people might not succeed in formal education. As Lori Kendall argues, even in the 1980s and 1990s, nerds-on-screen were still moving from a position of abject otherness to participation in hegemonic masculinity, often displacing and appropriating narratives of Black and gay characters.[11] However, the view expressed throughout *The Bell Curve*, that failure within the formal education system might be a sign of some psychological weakness that would lead people to commit property crimes, become welfare dependent, and give birth to children who would in turn become state-dependent appears closer to the book's message. One hundred years after Nordau, at the end of another century, fears of degeneracy had caught up with the ontologization of Intelligence as IQ.

In 1994, the Academy Award for best film was not won by *Quiz Show* but by *Forrest Gump*, a very different morality tale about a man who fell at an unusual point on the IQ distribution. According to the then speaker of the House or Representatives, Newt Gingrich, *Forrest Gump* was "a conservative film. People went to see it as a reaffirmation that the counterculture destroys human beings and basic values. Maybe being simple but good, and decent and romantic, is a lot better."[12] According to film theorist Thomas Byers, *Forrest Gump* was a film that both remembered history and emptied it out through a pastiche of past historical events stripped of meaning and reduced to their importance for the individual Forrest Gump, a man too stupid to perceive their relevance. There are many similarities between *Forrest Gump* and *The Bell Curve* as reconstructions of the history of the sexual revolution. Forrest's lost love Jenny comes unstuck when she takes up of the freedoms of the sexual revolution. Her character is also a pastiche, and she appears as a college student prepared to play the field, a *Playboy* model, a Vietnam War protester, a single mother, and a person with HIV. However, these various experiments in living only get her into trouble, and they send her running back to Forrest time and again when her choices put her in the way of harm. In contrast, Forrest's character is defined by his low intelligence and his inability to perceive both the broader historical relevance of events and his own individual sexuality. He doesn't know that his mother slept with the school principal in order to keep him from being placed in a special education class. He is the rational man of Terman's imagination; his first ejaculation—at least within the film's narrative—is a strange event to him, that he seems to barely understand it when it happens. As Byers notes, the film's moral terms of engagement are clearly valenced and gendered: "I'm not a smart man," Forrest tells Jenny, "but I know what love is"—suggesting that even stupid men know this, but promiscuous Jenny does not.[13]

The popularity of *Forrest Gump* tells us something about the zeitgeist of the early 1990s in which *The Bell Curve* became a best-seller. Both revisionist histories speak to the dangers of the sexual revolution, and both garnered poularity among academic and popular audiences. The popularity of these narrative revisions shows how partial Kinsey's normalization of sexual difference has been. However, these two narratives divergence utterly with regard to the value that they place on Intelligence to weather anxiety-provoking historical times. *The Bell Curve* suggests that one must have a certain IQ to survive a sexual revolution and to avoid being caught up in its collective fantasies. *Forrest Gump* tells us that the best way to live through history is to lack the worldly wisdom to notice that it is happening, and to allow only the most traditional fantasies about marriage and family to propel you forward. Jointly the narratives tell us that Americans who invest in high intelligence and those who place a value on home-spun wisdom shared anxieties about the normalization of sexual freedoms in this period.

The popularity of these two conservative revisionisms about the history of the sexual revolution might give us pause to wonder what was genuinely smart, at that time, about the enthusiasm for Foucault's critiques of sexual liberation. In the chapters above, I have aimed to make explicit some limits of the Catholic confessional as an adequate metaphor for the power/knowledge of the modern sexual sciences in the West. Unlike redemptive "confessional" practices of individual psychotherapy, Kinsey's sexology worked by inviting people to invest their trust in the possibility of a new science of sex. I meant the initiation metaphor to do three things: to foreground how the harnessing of collective investments in futures effects power; to signal the co-proliferation of multiple popular and academic sexologies and psychologies with characteristically low epistemological status in the modern world; and to make explicit how secular theories anchor critique in religious metaphors, and so tend to use past forms of understanding as implicit critique.

Certainly, the surveying of sexual behavior remained a heterogenous "low" science after Kinsey. It never became the idealized paradigmatic science organized by the logic of random sampling imagined by Cochran, Mosteller, and Tukey. Rather, after the publication of *Sexual Behavior in the Human Female*, Kinsey's research was de-funded by the Rockefeller Foundation. Staff at the Institute for Sex Research attempted a national sample once more, but their results went unpublished.[14] However, the mystique of the sex survey has been persistent. In the era of the sexual revolution, the practices of surveying sex were taken up largely in the interests of profit and of selling knowledge about the normality of sex to members of the public; a fractured market opened up for "sex surveys" based on reader's polls in publications such as *Psychology Today, Playboy*, and *Cosmopolitan*. For several decades after Kinsey, the genre of the sex survey advanced no further than the *Literary Digest*'s attempt to predict the 1936 presidential election. Disparate, corporate cultures of popular sexual knowledge characterized the United States, but division was more common than consensus.

As a result of this heterogeneity and uncertainty, the psychologist Dana Bramel's hypothesis, that people might defensively project statistics about matters such as homosexuality onto the nation, remains relevant to understanding the psychology of power/knowledge. We can now pick and choose the sex surveys by which we want to organize our intelligence of normality. In the 1970s, Bruce Voellmer of the National Gay Task Force was in search of a statistic by which to project gay people as a legitimate minority group into national debates. Voellmer (1990) selected all of the participants scoring above a certain point on Kinsey's scale (7 percent of women, 13 percent of men) to suggest that 10 percent of the population was lesbian or gay. This claim achieved wide currency and "looped" in Hacking's sense through gay and straight culture, and through the HIV/AIDS policies of the 1980s. Contrary to Kinsey's insistence in 1948 that one cannot separate the sheep from the goats, Voellmer turned the homosexual into a species again forty years later so that he could better engage a nation state utterly organized by the legitimacy of representativeness and ontologies of sampling.

In the 1990s, and as a result of a collective panic to purge the nation of the HIV superbug, a precarious alliance allowed a national sex survey (Laumann et al. 1994) to be conducted under the auspices of the NORC. The enrollment of bipartisan political support for such a survey affected the range of questions that were asked, and the ultimate result was a picture of Americans as far more conformist in their sexual behavior than Kinsey had painted.[15] As sexual politics moved onto geek-driven cyborg frontiers of inferential statistics and populations of varying probability, lesbian/gay organizations predictably cited the higher Kinsey figure, while homophobic organizations tended to cite the lower NORC figure.[16] As sex surveys not only demystify, but also require investment in new knowledge that might be garnered from such statistical training, Americans can still choose to believe very different facts about who makes up the "mythical norms" of the nation. Terman's homophobic investment in Kinsey's statistics about the frequency of homosexuality contrasts with contemporary homophobic investment in the projection of very few gay and lesbian people onto the population.[17] However, it is difficult to conclude that Dana Bramel's experiment on defensive projection does not remain culturally relevant.

Latour has argued that when concepts are translated to new domains, new networks are formed, and that the concepts also change in the process. Reworking Foucault's analysis of normalization to account for Kinsey and Terman led me to examine persistent fantasies about frontiers. Frontier fantasies have fanned the flames of enthusiasm for new forms of human science that might normalize atypical kinds of people: gifted children, rural bisexual men, imaginative masturbators, less-than-ideal husbands, taxonomists. Frontier fantasies have a singular future-oriented focus that is quite the opposite of the form of decade-slumming genealogy that I have been practicing in this book. But in its impulse to forget the past, frontierism becomes very vulnerable to reconstructions of that past that change the way we feel about possible versions of the future. Only if we forget that the frontier of Turner's imagination was already populated can we invest in fantasies about American frontiersmen. Only when we forget Freud's debts to Jewish thought does Freud appear as psychology's one man of genius. Only if we forget that children play clusters on the piano spontaneously are we impressed that Cowell "invented" them as a child. I am wary of frontier narratives that aim to inspire awe and wonder in the here and now for the forgetting that they urge.

At the end of Bill Condon's uplifting film *Kinsey*, viewers are left with an image of Clara and Alfred among the Californian redwoods, symbols of American nature in all its permanence and warmth, much loved by Californian eugenicists. The couple walk away as Alfred urges Clara on to more work. It's an uplifting image, rooted in American soil, marital happiness, and the urge to love and to work. But Alfred never wanted a Hollywood film made about him, and the film stops short of his early death, hastened by overwork. I want to end this book on a dissonant note about Alfred Kinsey's "nature" that cautions against the tireless investment of life in the human sciences, which ends Condon's film.

Consider instead the material "other" to both the national and species boundaries of Kinsey's science—the Old World European wasps. It is to those neglected and excluded beings that I devote my last words. Kinsey knew that the tannic acid in the galls produced by European wasps had for centuries been combined with iron to make the iron sulfate in "iron gall ink." Iron gall ink was a staple of the European logocentric tradition, but by the mid-twentieth century trade in iron gall ink had been long in decline. The original users of this substance could not have predicted how iron gall ink reacts slowly over centuries with trace metal elements in paper, slowly erasing the words, images, and musical notation into which it has been drawn.[18] Much like Freud's mystical writing pad and Derrida's theory of deconstruction, the letters, drawings, and notations made material by gall wasps' abandoned bedchambers are involved in an unfaithful process of writing and erasure. European gall wasps have a particular claim among insects to be "zoology's Other, the definitive organisms of *differance*."[19] The sublimation of metals in iron gall ink presents contemporary archivists with the horror of a world of copies of Western knowledge which lack originals. The biology of the European gall wasps was sidelined by Kinsey's science, but the energy extracted from the bedchambers in their post-natural relationships with humans remains unstable. At the horizons of understanding, at the molecular level of writing, the debts that writing owes to insect labor are becoming unstable as debts accrue. The world remains naturally unpredictable, even if knowledge sometimes initiates humans into the presentist collective fantasy that we can predict and control the future on the basis of stable knowledge about what has happened, what normally happens, and what we should aim to have happen in the future.

Notes

Abbreviations for Archival Sources

KP: Alfred C. Kinsey Papers, Archives of the Institute for Sex Research, Indiana University

SP: May Seagoe Papers, Archives of the History of American Psychology, University of Akron, Ohio

TP: Lewis M. Terman Papers, Stanford University Libraries

YP: Robert Mearns Yerkes Papers, Manuscripts and Archives, Yale University Library

Chapter One

1. *SBHM*, 623; on the origins of the ban of gay men and lesbians in the service, see Bérubé 1990; on the pathologization of homosexuality by the American Psychiatric Association's Diagnostic and Statistical Manual, see Bayer 1981.
2. *SBHM*, 671; The United Nations, *The Universal Declaration of Human Rights*, adopted by the United Nations on December 10, 1948.
3. T. C. Boyle, *The Inner Circle* (2004); Bill Condon, director, *Kinsey* (2004); Barak Goodman and John Maggio, directors, *American Experience: Kinsey* (2005).
4. Lyle Lanier, Letter to Lewis M. Terman, May 18, 1948. TP.
5. Lewis M. Terman, Letter to Lyle Lanier, June 8, 1948. TP.
6. On Terman's life and work, see Chapman 1988; Minton 1988; Seagoe 1975; Terman 1930.
7. On the mass testing of American soldiers during World War I, see Carson 1993; Hornstein 1988; Kevles 1968.
8. The exchange between Terman and Lippman is reproduced in full in Block and Dworkin 1976, 4–44.
9. Smuts 2006.
10. See Minton 1988, 91–95; Seagoe 1975, 64–79.
11. The principal reports of this research were contained in four volumes of

Terman's "Genetic Studies of Genius" book series; see Terman 1925; 1930; 1947; 1959.

12. Contemporary uses of Terman's data on gifted children by psychologists include Friedman, Kern, and Reynolds 2010; Lippa, Martin, and Friedman 2000; McCullough, Tsang, and Brion 2003. Critical accounts of Terman's construction of gifted children include Cravens 1992; Hegarty 2007; Margolin 1994; Shurkin 1992.

13. In contrast, research forwarding the idea that IQ was determined by environmental factors was promoted by researchers at the Iowa Child Welfare Station. For histories of the research at Iowa, see Cravens 1993 and Minton 1984.

14. Privateer 2005, 3–8.

15. Boring 1923. The issue surfaced again in the 1990s as psychologists struggled to define "intelligence" in the wake of the popular conservative volume, Herrnstein and Murray's *The Bell Curve* (1994). See particularly Neisser et al. 1996. One of the ways in which this statement was prematurely confident about what was "known" about intelligence was its claim that the tests were relatively immune to nonverbal forms of cultural bias. "Stereotype threat" was simultaneously posited as a process by which individual members of social groups are made aware of stereotypes linking group identity and test performance, and test performance is correspondingly impaired. See Steele and Aronson 1995. For a more recent review of stereotype threat research, see Schmader, Johns, and Forbes 2008. For a more recent statement of the position that intelligence is only what IQ tests measure, see Richardson 2002.

16. Lewis M. Terman, Letter to D. D. Klein, Oct 26, 1948. Box 17, Folder 31, TP.

17. This description of Terman is offered by Jones 1997, 588.

18. See Gathorne-Hardy 1998, 275–76; Jones 1997, 588–92; Minton 1988, 227–32; Pomeroy 1972, 290–94; Seagoe 1975, 147–48.

19. On Carney Landis's research and its relationship to the history of sex research in psychology, see Serlin 2012.

20. Jones 1997, 590–91.

21. On the life and work of psychologist Robert M. Yerkes, see Dewsbury 2006; Haraway 1989; Pickren 1997; Yerkes 1932.

22. The CRPS funded research that Terman began in 1925 on the measurement of masculinity-femininity as an aspect of human personality, research on the definition and measurement of marital happiness, and a study of the marital happiness of his gifted cohort in adulthood; see Terman 1938; Terman, Bayley, Marshall, McNemar, and Oden 1951; Terman and Miles 1936. These research projects are discussed in greater detail in chapters 3 and 4 below.

23. For a history of the Committee for Research on Problems of Sex, see Aberle and Corner 1953; Pickren 1997. On the avoidance of controversy at the CRPS, see Clarke 1990 and 1998.

24. Robert M. Yerkes, Letter to George W. Corner, December 30, 1948. Box 12, Folder 203. YP.

25. See Clarke 1998, 116.

26. Kinsey was first granted $1,600 in 1941, but by the 1946–47 fiscal year was being granted $35,000 per annum from the CRPS; Aberle and Corner 1953, 121.

27. The literature on American eugenics is vast. I have relied particularly upon Kevles 1985. On California, see Kline 2001 and Stern 2005. On international exchanges between U.S. eugenicists and German Nazi eugenicists, see Kühl 2001. For a discussion of the continuation of eugenic sterilization in the South, see Roberts 1997. The human genome project prompted further historical reflection on American eugenics in the 1990s. See Kevles and Hood 1993; Nelkin and Lindee 1995; Ordover 2003.

28. "The more a body of knowledge is understood to be disinterested and objective, the more valuable it is as a tool in moral and political action. Conversely, the capacity of a body of knowledge to make valuable contributions to moral and political problems flows from an understanding that it was not produced and evaluated to further particular human interests" (Shapin 1996, 196).

29. On discourses about Irish Intelligence, see Benson 1987.

30. See D'Emilio 1993.

31. On the discourse of "political correctness" in the 1990s, see particularly Fairclough 2003.

32. Among the feminist and queer writings on science that influenced me most were Butler 1993; Keller 1985; Haraway 1991; Harding 1987.

33. For accounts of this work, see Hegarty 2001; Hegarty and Pratto 2001.

34. Among the more influential texts of the discursive turn in social psychology, both in the field in general, and on me in particular, were Billig 1987; Burman and Parker 1993; Potter and Wetherell 1987.

35. Kuhn 1970. For an early application of Kuhn to psychology, see Buss 1978. For a cogent critique of the influence of psychologists on Kuhn and of Kuhn on psychologists, see Driver-Linn 2003.

36. I borrow this phrase from Cravens 1992.

37. Gergen 1985.

38. See particularly Danziger 1990; Gigerenzer et al. 1989; Hacking 1990; Morawski, ed., 1988.

39. See particularly Danziger 1991; Krantz 2001; Morawski 1994; Richards 2010; Smith 1988. On the differences between historians' narratives of the past, the evidence of the past, and past events, see also White 1973 and 1987.

40. See Boring 1929.

41. Hegarty 2001.

42. My metaphor of the past as a contested political zone borrows from Donna Haraway's account (1991) of the gender politics that work through narrating the past history of primatology.

43. On the genre of scientific psychology, see Budge and Katz 1995; Walsh-Bowers 1999.

44. The attempts of experimental social psychologists to narrate the history of their field have particularly interested historians. See, e.g., Cherry 1995; Samelson 2000; Lubek 2000; Moghaddam and Lee 2006.

45. See particularly Bailey and Zucker 1995; Daryl J. Bem 1996.

46. Hegarty 2003.
47. Foucault 1978. See also Katz 1976; Smith-Rosenberg 1985; Weeks 1977.
48. See particularly Henriques et al. 1984.
49. Henriques et al. 1984.
50. Danziger 1997. For an insightful critique of psychologists'readings of Foucault's theory of discourse, see Hook 2001.
51. Hartsock 1990.
52. Hegarty 2007.
53. See Halperin 1995.
54. I borrow this phrase from historian of psychology Frances Cherry (1995), who uses the unexplained "particulars" of social psychology experiments' methods and results to position those experiments as historical.
55. Foucault 1977a and 1978. For a cogent account of the relevance of genealogy to American sex research, see Terry 1999, 20–21.
56. It is important to note that Foucault's writings on discourse have been taken to be a "method" of discourse analysis that can be applied as a form of psychology. Within psychology, the term "Foucaultian discourse analysis" often applies to the analysis of hidden assumptions about forms of power such as patriarchy and racism at work in talk and text. Such analyses typically acknowledge the historically situated nature of social discourses but do little to engage with historical change in discourse. For a Foucaultian critique of this shortcoming of "Foucaultian discourse analysis," see Hook 2001.
57. Foucault's concept of the *dispositive* extends beyond the linguistic to material forms of power. See for example the development of this idea in the engagement between Foucault and feminist psychology in Amigot and Pujal 2009.
58. Several studies have drawn attention to the distinct ways that intelligence testing has developed in different national contexts to fit with, and to reinforce, distinct national ideologies about equality, inequality, and civic participation. See, for example, Carson 2007; Mulder and Heyting 1998; Ramsden 2007; Richards 2004; Schneider 1992; John White 2006.
59. See for example Carson 1993; Margolin 1993; Rose 1996.
60. By differentiating "Intelligence" (with a big I) from "intelligence" (with a small i), I mean to draw on Graham Richards (2008) argument that the social construction of that body of knowledge we call "Psychology" must proceed through routes that are at least somewhat "psychological." Like Richards, my use of two different terms for "I/intelligence" is intended to throw into relief the assumptions about the "intelligence" needed to research "Intelligence" successfully, as well as models of the intellect that are displaced or that compete with the Intelligence that IQ tests measures as a means of talking about intellectual differences between people. This distinction is particularly salient in chapter 3 below.
61. See particularly Daston 1987.
62. Darwin 1859.
63. My thinking on these two models of normativity borrows heavily from Gigerenzer et al. 1989 and Hacking 1990.
64. Kurt Danziger (1990, 49–67) takes Galton's anthrometric measurements at

the 1884 International Health Exhibition in London to be foundational to the measurement of individual differences in modern psychology.

65. Cattell 1890. Cattell had developed in interest in the use of tests to study individual differences while a student of Wundt's at Leipzig. His interest was discouraged by Wundt but later encouraged by Francis Galton. On the development of Cattell's ideas, see Carson 2007, 172–77, and Fancher 1985.

66. See the entries on the Darwin family in Galton 1869, 209–10; 1874, 47–48. The latter reads: "Among those who do not bear that name (being children of the daughters of Dr. Erasmus Darwin), I mention (3) myself, with all humility, as falling within the limits of the group of scientific men under discussion, on the ground of former geographical work, and have had much to do in the administration of various scientific societies."

67. See for example Simonton 1999.

68. Becker 1978.

69. Huertas 1993.

70. On the increasing adherence of abnormality to sexuality in the late nineteenth century, see Foucault 2003.

71. Freud also used this strategy in a letter to an anxious American mother who hoped for a psychoanalytic "cure" for her son's homosexuality. Freud reminded her that "many highly respectable individuals of ancient and modern times have been homosexuals, several of the greatest among them (Plato, Michelangelo, Leonardo da Vinci, etc.)" and advised her to read the works of Havelock Ellis. The letter was sent to Alfred Kinsey and has been framed and displayed in the Kinsey library. See Abelove 1993.

72. Freud, "Three Essays on the Theory of Sexuality," SE 7:156, 178–79, 238.

73. Freud, "Civilized Sexual Morality and Modern Nervous Illness," SE 9:190.

74. Freud, "Leonardo Da Vinci," SE 11:132–36.

75. Late-nineteenth-century theorists who vilified genius—such as Royce (1891, 18–19)—and those who valorized it—such as Galton (1869)—considered the inherited gifts of the father to be more relevant than those of the mother in accounting for the emergence of genius.

76. While this quotation is often attributed to Lombroso, Lombroso (1889, 138) attributed it originally to Goncourt.

77. Freud, "Civilized Sexual Morality," SE 9:195. This theme returns in Freud's later essays on women, where he characterizes men as having both a weaker superego, as a result of the Oedipus complex's formation after castration, and more sublimated "libidinal cathexes," compared to women. Relatedly, Freud described the capacity of a woman to carry on an intellectual profession as something that might be "recognized as a sublimated modification" of penis envy. See Freud, "Some Psychical Consequences of the Anatomical Distinction between the Sexes," SE 19. On the conflation of genius and maleness see Battersby 1989.

78. Gibson 1998.

79. For an internalist critique of Terman's account of the gifted women in the study, see Sears and Barbee 1977.

80. Horney 1926.

81. Mauss [1924] 2002, 80–81.

82. For a critique of the inclusivity of Terman's gifted study, see Cravens 1992. On the over-representation of White children in giftedness programs, see Margolin 1993.

83. In chapters 4 and 6 below I consider historical claims about the relationship between genius and homosexuality and Jewishness in greater detail. For a history of African-American psychologists contribution to the discipline, see Guthrie 1976. On the contributions of American women psychologists, see Furumoto and Scarborough 1986; Johnston and Johnson 2008; O'Connell and Russo 1990. On genius, gender, and sexuality, see Elfenbein 1998 and Battersby 1989.

84. For example, histories of early women psychologists have relied on James McKeen Cattell's work (1906) to determine which women psychologists are worthy of historical attention. See, for example, Furumoto and Scarborough 1986 and Johnston and Johnson 2008.

85. See Stewart 2003.

86. Carson 2007. Similarly, Graham Richards (2004) argues that the connection between intelligence testing and scientific racism was very much "an American thing."

87. See also Chapman 1988.

88. On the continuity between nineteenth-century common sense philosophy and twentieth-century American psychology, see Carson 2007 and Richards 1995.

89. Terman 1941. See also Seagoe 1975, 140.

90. See Terman 1930.

91. Seagoe 1975, 182. For a contemporary argument in favor of using historical sources to estimate the IQ, see Simonton 2009.

92. See Pomeroy 1972, 14.

93. See ibid., 431–32.

94. See Christenson 1971 and Pomeroy 1972.

95. For a cogent comparative analysis of Kinsey biographies, see Capshew et al. 2003.

Chapter Two

1. This is not to say that his female participants were not subversively smart; many had learned to masturbate from literature aimed at its suppression. *SBHF*, 139.

2. Terman 1948, 452. Kinsey claimed that about two-thirds of boys' first orgasms occurred through masturbation.

3. In the summer of 2010 I conducted a search using the database "Psycinfo" for articles that included the word "masturbation" in the record, yielding 275 articles. Of these, only 64 were published in English in the twentieth century prior to 1970. Among them precious few were published in journals that were not medical, psychiatric, or psychoanalytic in orientation. A notable exception to this trend is a pre-Kinsey survey of college men (Pullias 1937), which found that over 80 percent believed that masturbation had some sort of serious damaging effects, and whose author raised concerns about the fears and guilt created in young men by anti-masturbatory instruction.

4. On the history of Enlightenment thought about masturbation, see Stengers and Van Neck [1984] 2001; Laqueur 2003; Neumann 1975.
5. For an important analysis of the relationship between masculinism and lack of reflexivity in psychological science, see Morawski 1994.
6. Danziger 1990.
7. Morawksi 1992; see also Morawski 2005.
8. Prior to Terman's appointment as its chairman in 1921, the department had been led by the introspectionist Angell, and had produced little work of enduring impact. Under Terman, the department rapidly garnered several grants to develop tests, built a network of researchers around that work, long before Stanford became such a powerful institution of research. See Lowen 1997 and O'Donnell 1985.
9. See Laqueur 2003, 51, 368–70; Stengers and Van Neck [1984] 2001, 138.
10. For critical discussions of Hall's account of masturbation within the history of psychology, see Arnett 2006, 186–97, and Cravens 2006.
11. Bertolini 1996.
12. Rosenman 2003.
13. Bullough 1987.
14. On the pedagogical environment at Clarke under Hall, see Sokal 1990, 114–24. On the influence of common sense philosophy on G. Stanley Hall, see Sheldon H. White 1992.
15. I rely primarily here on Ross 1972.
16. Harper 1949.
17. Rosenzweig 1992.
18. Ross 1972, xiv.
19. See particularly Hall 1885.
20. Ross 1972, 254.
21. On the gendered division of labor in Hall's child study movement, see Ehrenreich and English 1979; Hoogland 2004. On Hall's view of the intellectual capabilities of women and men and segregated education, see Diehl 1986.
22. See Ross 1972, 341–67.
23. See Freedman and D'Emilio 1992; Rosenberg 1985; Maines 2001.
24. For a more extensive discussion of race in Hall's work, see Muschinske 1977.
25. Diehl 1986.
26. As such, Hall's work fell within a long line of romantic thought in which fears about human weakness and limitless appetites for masturbation are heightened by the seemingly endless powers of the human imagination. See Laqueur 2003. Rousseau described masturbation as a threat to education: "It would be a dangerous matter if instinct taught your pupil to abuse his senses; if once he acquires this dangerous habit he is ruined . . . body and soul will be enervated; he will carry to the grave the sad effects of this habit" (Rousseau 1963, 180).
27. For example, Hall considered some of the worst pronouncements of Victorian medicine to be overstated: "In discussing the results of onanism, we must first of all recognize that the immediate and sensational effects often seriously believed in, and often purposely exaggerated for pedagogical effects, are not so immediate or disastrous as represented in both the popular

and the earlier literature. The brain is not literally drained away; dementia, idiocy, palsy, and sudden death are not imminent, nor is there any peculiar infallible expression, attitude or any other manifestation instantly recognizable by experts" (1904, 439).

28. See Diehl 1986.

29. Sumner's dissertation on the work of Freud and Adler was the last that Hall would supervise. Sumner later led the training of generations of African-American psychologists as head of the psychology department at Howard University. However, Sumner's writings often reiterated the view that Black folk were less "culturally advanced" and suited to more basic education than were whites. On the life and work of Cecil Sumner, see Guthrie 2000 and Sawyer 2000.

30. See Ott 1996. On the cultural history of tuberculosis in the late nineteenth century, see also Bates 1992; Rothman 1994; Teller 1988.

31. Sokal 1990.

32. For all of the rhetoric of physical health that organizes Hall's genetic psychology, he rarely mentioned tuberculosis—the leading cause of death in the United States in the nineteenth century—in *Adolescence*. In the one place where he did (1904, 257–58), he applied his compensatory logic to the disease, arguing that it was particularly likely to strike the body during growth spurts when the nutritive powers were exhausted.

33. Terman's tuberculosis returned particularly in 1904, 1928, and 1936 (Seagoe 1975, 16).

34. Terman described his experience of Hall's Monday night seminar as surpassing all of his other educational experiences (Minton 1988, 23).

35. The most effective critical response to Cattell's attempt to measure intelligence was Clark Wissler, "The Correlation of Mental and Physical Tests" (1901). As Graham Richards notes (1997, 32–33), the conceptual relationship between reaction time and intelligence was inconsistent in the psychological literature of the 1890s. While Galton and Cattell assumed that faster reaction times indicated higher intelligence, those studies which found Black participants to be quicker than White participants interpreted the race differences as evidence of the primitivism of the Black participants.

36. Spearman 1904. Spearman understood that Cattell's earlier observation of uncorrelated performance on tasks had led to the conclusion that there was no general intelligence because of the error of measuring the ability with each test. See Spearman 1930.

37. Binet and Simon 1905. On the life and work of Alfred Binet, see Fancher 1998; Foschi and Cicciola 2006; Siegler 1992.

38. Freud, "Three Essays on the Theory of Sexuality," *SE* 7:125–245.

39. This empirical study was conducted under the supervision of the experimental psychologist Edmund Clark Sanford. Sanford had been a student of Hall's, and became Hall's first faculty appointment to the psychology department at Clark. Sanford has been best remembered for his laboratory manual. On his life and work, see Goodwin 1987.

40. Buenker and Kantowitz 1988.

41. Kurt Danziger (1997, 66) notes the disparity between the omission of "intelligence" from Baldwin's *Dictionary of Psychology* at the start of the twentieth

century and its wide currency as an academic and popular term in the early 1920s, to describe possibly measurable differences between people's intellectual capacity. On the development of intelligence testing in the United States, see also Carson 2007; JoAnne Brown 1992; Fancher 1985; Gould 1981; Guthrie 1976; Privateer 2006; Richards 1997; Chapman 1988; Zenderland 1998.

42. Goddard 1910. On Goddard's life and work, see Zenderland 1998.

43. Gould 1981, 177.

44. Fancher 1985, 139.

45. Carson 2007, 183.

46. Several historians have noted how technically "incorrect"—but clever—responses to IQ tests were miscategorized as simply "incorrect," conflating conformity with intelligence. See Carson 2007, 189; Gould 1981, 175–76; Guthrie 1976, 52. While Terman considered the Intelligence that his test measured to be inherited, he admitted that "no adequate definition can possibly be framed which is not based primarily on the symptoms empirically brought to light by the test method" (Terman 1916, 44).

47. JoAnne Brown 1992.

48. On the social hygiene movement, see Brandt 1985 and Moran 2000.

49. Lunbeck 1995.

50. See Kline 2001, 40–42.

51. Carson 2007, 192–93; see Margolin 1993.

52. Minton 1988, 111.

53. See Terman 1925, 19–25, for a detailed description of the recruitment procedures in this study. Other more critical accounts are reported in Cravens 1992; Minton 1988, 112–17; Shurkin 1992, 30–36. Unlike some of his contemporaries who studied high-IQ children, Terman failed to recognize the possibility that children could be young, gifted, and Black. In contrast, psychologist Leta Stetter Hollingworth set up a program for gifted children in New York City that included Black children. On her life and work, see Benjamin and Shields 1990 and Shields 1991.

54. Dorothy Yates, Letter to Lewis M. Terman, January 2, 1922. TP.

55. Helen Marshall, Letter to Lewis M. Terman, October 22, 1921. TP.

56. On situated knowledges of women researchers, see Haraway 1991 and Harding 1987. On the capacity of IQ tests to assess the intelligence of African-American subjects, see Bond 1924 and 1926. On Bond's life and work, see Urban 1989. For a discussion of Bond's marginality as a source of reflexive insight, see Morawski 2005.

57. Terman 1925, 173–214, for a description of these studies.

58. *SBHM*, 298; Moran 2000, 68–97; Stengers and Van Neck [1984] 2001, 123–35.

59. For a review of the initial critical reaction to Jones's biography, see Gathorne-Hardy 1998, 363–67.

Chapter Three

1. Alfred C. Kinsey, Letter to Lewis M.Terman, September 27, 1946. KP.

2. On the life and work of Quinn McNemar, see Hastorf, Hilgard, and Sears 1988; McNemar 1980.

3. Alfred C. Kinsey, Letter to Lewis M. Terman, September 27, 1946. KP.

4. Lewis M. Terman, Letter to Alfred C. Kinsey, October 2, 1946. Kinsey Institute.

5. Stern 2005, 161. On the history of the AIFR, see also Kline 2002.

6. Stern 2005, 552, 556.

7. See Gebhard and Johnson 1979, 28–29.

8. Alfred C. Kinsey, Letter to Lewis M. Terman, January 15, 1952. KP. Terman's paper was published as Terman et al. 1951.

9. See, for example, Scully and Bart 1973.

10. Neuhaus 2000, 447–73.

11. For examples of social histories that cite Kinsey's statistics as realist evidence of this historical shift, see Brandt 1985, 175; Gordon 1977, 193; Moran 2000, 79, 96, 136; Reed 1983, 61.

12. See, for example, Gottman and Notarius 2002.

13. On the life and work of E. Lowell Kelly, see Anon. 1986; Fiske, Conley, and Goldberg 1987.

14. E. Lowell Kelly, Letter to May V. Seagoe, September 22, 1967. AHAP.

15. Quoted in Chesler 1992, 217. On Popenoe's attitude to Sanger, see Reed 1983, 134–35; Ordover 2003, 141. On Popenoe's eugenic activism and post-war career as the resident marriage expert in *Ladies Home Journal*, see Kline 2001; Ladd-Taylor 2001; Stern 2005.

16. Stern 2005, 79.

17. Kline 2001, 50–55.

18. On Popenoe's support for Hitler and the German Nazi regime after 1933, see particularly Popenoe 1934; Kline 2001; Kühl 1994; Ladd-Taylor 2001; Stern 2005.

19. Minton 1988, 147–50.

20. Stern 2005, 174.

21. Lewis Terman, Letter to Robert M. Yerkes, January 29, 1951. TP.

22. Popenoe described interracial marriage as biologically unsuitable, whether "one race is markedly inferior to the other," as he assumed to be the case in "a Negro-white marriage," or "in which the races are equally superior, but in markedly different ways—Japanese and Americans, for instance" (Popenoe 1925, 47).

23. When gynecologist Robert Dickinson developed less-painful methods of sterilizing women in California with Popenoe and Gosney's support, he proudly announced to his colleagues at the meeting of the American Medical Association that the eugenically sterilized woman was "sterilized, but not unsexed" (see Kline 2001, 67–78).

24. "The experience of the most intelligent people, for unnumbered years, has been that the normal sex life enjoyed in marriage is a source of the greatest power and satisfaction; that much of the finer side of existence is based on it, that it brings happiness, health, and the ability to do the work of the world to the best advantage. The men and women who have done the most good in the world have, on the whole, been those with full and rich love-lives" (Popenoe 1925, 136–37).

25. Funding patterns of the CRPS during this period are described by Terry 1999, 126–42; Ericksen 1998, 30–35, 38–41; Pickren 1997. While the CRPS

did not officially support Hamilton's research, members arranged for its sponsorship by the Bureau of Social Hygiene. See Clarke 1998, 103.

26. James Jones describes *Ideal Marriage* as "the best selling marriage manual in the United States" until 1969 (Jones 1997, 293).

27. Nicholson 2003, 188–89. On the rising importance of character and adjustment in American psychology in the 1930s, see Danziger 1997; Lombardo and Forschi 2003; Napoli 1980; Wake 2011.

28. See Burgess and Cottrell 1939. On Burgess life and work, see Bogue 1974. For a discussion of Burgess and Cottrell's work in relationship to Terman's, see Ericksen 1998, 41–47.

29. Minton 1988, 179.

30. E. Lowell Kelly, Letter to May V. Seagoe, September 22, 1967. SP.

31. Lewis Terman, Letter to Robert M. Yerkes, January 29, 1951. TP.

32. Donald P. Wilson, Letters to Lewis M. Terman, November 15, 1934; November 19, 1934; November 21, 1934; February 18, 1935; and March 9, 1935. Lewis M. Terman, Letters to Donald P. Wilson, December 3, 1934; February 5, 1935; February 27, 1935. TP.

33. Donald P. Wilson, Letter to Lewis M. Terman, December 21, 1934.

34. In contrast to this uncertainty about dating the first occurrence of masturbation, Terman was confident that first menstruation dated adolescence precisely among girls (*PFMH*, 253).

35. In spite of this instruction to the participants, Terman elsewhere described orgasm as an experience that "can be reported with more objectivity and accuracy than many of the other kinds of information which our subjects have been asked to supply" (*PFMH*, 304).

36. Terman, *PFMH*, 146; see Minton 1988, 182.

37. Terman's prioritizing of personality differences rather than differences in sexual experiences in this appendix is also at odds with the relative *size* of the statistical differences reported. Terman reports seven differences in the responses of the two groups of husbands on sexuality-related items, and eleven differences in personality items. The magnitudes of the differences are expressed in terms of Critical Ratios and are somewhat larger for the sexuality measures (average = 2.5) than for the personality measures (average = 1.7). *PFMH*, 403.

38. Aberle and Corner 1953, 129. The findings of this work are reported in Terman and Oden 1947.

39. See McNemar and Terman 1936.

40. Kelly's troubles during his first marriage affected not only his interest in marital research but also his personal relationship with the Terman family. See Hegarty 2007.

41. Jorgensen and Jorgensen 1999, 156.

42. Dorfman, 1984, 496.

43. See, e.g., Tilman 1996, 41. On Ellen Veblen and her treatment within the literature on Thorstein Veblen, see Eby 1999.

44. Fred Terman, Special Comments. Box M904 Folder 4. SP.

45. May Seagoe, Letter to E. G. Boring, January 10, 1968. Box M900 Folder 1. SP.

46. May Seagoe, Letter to Harry Harlow, July 10, 1967. SP.

47. See Blum 2002.

48. Harry Harlow, Letter to May Seagoe, July 17, 1967. SP.

49. By heteronormativity, I mean that ideology that presumes that heterosexuality both is and ought to be the universal condition. This critical term names a way of thinking that has sneaked into the most basic of categories and which presents itself not as an ideology but "clothes itself in goodwill and intelligence" (Warner, 1993, xxiii).

50. Jones attributes this difficulty to an "adherent clitoris" that was freed by dissection (1997, 236). Gathorne-Hardy queries this interpretation on the grounds that "an adherent clitoris doesn't prevent intercourse" (1998, 60).

51. On the incompatibility between the subject positions "gay," "lesbian," and "bisexual" and "behavioral scientist" in Terman and Kinsey's time—and in their networks—see Minton 2002. In the "lesbian-affirmative" and "gay-affirmative" genres of psychology made possible through the depathologizing of homosexuality as a mental illness, claims of expertise on the basis of "insider" status became more widely acceptable. See Kitzinger 1987, 29–31.

52. For other writings that encourage readers to enjoy stories in which gifted children, and the scientists who admire them, fail to make good on the promise of assumed natural superiority, see Cravens 1992; Plotz, 2005.

Chapter Four

1. When Kinsey received the copy of the review that Terman sent him, he wrote the word "moral" in the margin (Pomeroy 1972, 290).

2. Sedgwick 1990.

3. See Jones 1997, 589.

4. Indeed, Yerkes commented on the greater sales of SBHM over Terman's latest book on the gifted in a letter written while Terman was writing his review of Kinsey. Robert M. Yerkes, Letter to Lewis M. Terman, March 2, 1948. YP.

5. Lewis M. Terman, Letter to Alfred C. Kinsey, July 25, 1941. TP.

6. Alfred C. Kinsey, Letter to Lewis M. Terman, August 23, 1941. KP.

7. Lewis M. Terman, Letter to Alfred C. Kinsey, August 30, 1941. TP.

8. Lewis M. Terman, Letter to Alfred C. Kinsey, October 23, 1941. KP.

9. Lewis M. Terman, Letter to Robert M. Yerkes, March 11, 1953. TP.

10. Ross 1950. This short report argued that similar instances of sexual behaviors—masturbation, homosexuality and premarital intercourse—were found using anonymous questionnaires and Kinsey's interview method.

11. Sedgwick's list includes "the categories secrecy/disclosure, knowledge/ignorance, private/public, masculine/feminine, majority/minority, innocence/initiation, natural/artificial, new/old, discipline/terrorism, canonic/noncanonic, wholeness/decadence, urbane/provincial, domestic/foreign, health/illness, same/different, active/passive, in/out, cognition/paranoia, art/kitsch, utopia/apocalypse, sincerity/sentimentality, and voluntary/addiction" (1990, 11).

12. Kinsey et al. also describe the research on human sexual behavior as born out of his earlier field research on gall wasps (SBHM, 9).

13. For related arguments about the influence of the discipline of biology on

Kinsey's sex research, see Pauly 2002; Rosenzweig and Rosenzweig 1969; Weinrich, 1990. While these perspectives are oriented toward explaining features of Kinsey's sex research, Kinsey's work in entomology is placed most firmly within the evolutionary work of its time by Kohler 2002.

14. In his criticism of Kinsey, Gould refers to the new systematics of Ernst Mayr to support his argument that Kinsey ought to have used this criterion to define the concept of the species. However, Mayr's own history of taxonomic understandings of the species problem shows that this idea had a pre-Darwinian precedent in the work of Buffon, and that there are contexts in which it ought not to apply (Mayr 1942 and 1982). Kinsey was aware of Mayr's 1942 work and cited it as a development in modern taxomony (*SBHM*, 17).

15. On the place of social and asocial insects in the social imaginations of humans, see the essays collected in Eric C. Brown, *Insect Poetics* (2006); Rogers 1989; Russell 1996.

16. On banal nationalism, see Billig 1995.

17. See Sleigh 2007, 2–6.

18. Russell 1996.

19. See Stipanovic 2006 and Olbricht 2006.

20. Sleigh 2007 and 2002.

21. On Wheeler's influence on Kinsey, see Christenson 1972, 32–34; Gathorne-Hardy 1998, 42; Jones 1997, 132–37.

22. Sleigh 2007, 58.

23. Gathorne-Hardy 1998, 48.

24. Jones 1997, 142.

25. See Faragher 1994.

26. Kinsey further notes (1937, 8) that Darwin was ignorant of the relevance of intergrades to evolutionary change.

27. On the attempts of military psychiatrists to screen the troops for gay men during World War II, see Bérubé 1990.

28. Minton 1988, 229; Lewis M. Terman, Letter to Robert M. Yerkes, December 8, 1955. TP.

29. See Minton 1988, 177. Aaron Rosanoff, a psychiatrist who gathered homosexual men's scores for Terman, also took an M-F test himself to allay his own fears—implanted earlier by the psychiatrist William Alanson White—that he might himself be a latent homosexual (Kline 2002, 136).

30. Lewis M. Terman, Letter to Robert M. Yerkes, December 8, 1955. TP.

31. This matter of whether psychological properties—such as intelligence and personality traits—could be measured in the absence of meaningful zero points for those measurements was explicitly debated earlier by Terman and Walter Lippman when the IQ concept became popularized. See Block and Dworkin 1976, 4–44.

32. Aberle and Corner 1953, 129.

33. Cox's volume was the only one of the *GSG* series that was not an empirical study of the gifted cohort.

34. Lewis M. Terman, Letter to Catharine Cox, March 2, 1927. TP. On the careers of women psychologists of Miles's generation, see Johnston and Johnson 2008.

35. On Barbara Burks's life and work, see King, Montanez-Ramirez, and Wertheimer 1996.
36. On the emergence of the category of the lesbian in the United States, see Duggan 2000 and Faderman 1991. See also the biography of Jeanette Howard Foster (Passet 2008), who worked for a time as Alfred Kinsey's librarian.
37. For the case history of "Roberta," see Terman Burks, and Jensen 1930, 330–31.
38. For the case history of "Renwick," see ibid., 328–29.
39. On the notion that gender is "in the making" in the interdependence of the organization of scientific work and the literary outputs of such work, see Haraway 1997.
40. Minton 1988, 171.
41. Terman and Miles 1936, 60.
42. Minton 1988, 167; Seagoe 1975, 142.
43. Indeed, the criminologist Norman Fenton was using the test for precisely these purposes in the Whittier school for juveniles even prior to the publication of *Sex and Personality*. See Hegarty 2007.
44. The test developed by Edward Strong showed working-class men to be more masculine than middle-class men, leading him to conclude that the "essential difference" between his own test and Terman's was that "education and presumably intelligence influence their MF scores to a greater degree than our MF scores" (Strong, 1948, 236). Catharine Cox Miles contributed to both Strong and Terman's concurrent attempts to measure gender as a personality trait. Other researchers found that Terman and Miles's test was inapplicable to the kind of forensic goals that Terman imagined for it. For example, one researcher (Walker 1941) concluded that only a minority of prisoners in Indiana prisons had sufficient intellectuality to complete its items.
45. Terman and Miles 1936, 13–14.
46. Terman and his colleagues' inability to think of feminine-typed behaviors performed by people-gendered-male as exemplary of intelligence is not, of course, determined by the nature of "intelligence," "gender-typing," or "people-gendered-male." Nor is it determined by their orientation toward quantification. An impactful contrasting case of the definition of intelligence can be found in the foundational essay on machine intelligence by gay computer scientist Alan Turing. Turing (1950) took the successful imitation of a woman by a man as the exemplary kind of human behavior that machines must successfully imitate before they can be attributed "thinking." For a useful exploration of the implications of Turing's test for queer theory, see Curtain 1997.
47. Lewis M. Terman to Robert M. Yerkes, February 5, 1935. Folder 909, Box 46. YP.
48. Terman, Miles, and Yerkes were all invited to join the Committee for Sex Variants in 1935. Organized by the gynecologist R. L. Dickinson, this committee conducted in-depth studies of homosexual subculture in New York. After 1935, the work was headed by George Henry and developed largely

by two unacknowledged gay researchers, Jan Gay and Thomas Painter. The findings are contained in Henry 1941, and provide unique insights into that culture at that time in spite of Henry's own normalizing arguments that homosexuality. For discussion of the crafting of this work, see Terry 1999, 220–67, and Minton 2002.

49. I describe these dynamics between Terman and Miles in greater detail in Hegarty 2012.

50. Quoted in Hicks 2005, 56.

51. Ibid., 77.

52. Quoted in Hicks 1993, 439.

53. On this problem see Shurkin 1992 and Cravens 1992.

54. See Hubbs 2004.

55. Terman is here quoting the opinion of Paul Rosenfeld in *The Dial*, in 1924. Rosenfeld's review is reprinted in full in Hicks 2005, 151–52.

56. Hicks, *Henry Cowell*, 235.

57. Cowell, "Process," 236.

58. A more skeptical assessment of Cowell's "string piano" technique, which involved playing the strings of the piano directly, was offered by pianist Olga Samaroff. "I have been experimenting with the 'string piano' without suspecting the magnitude of what I was doing, since I was five years old, and I am sure of the melancholy fact that I am older than Mr. Cowell. I have played upon it with clothes brushes, dust clothes and hairpins. I have a specially developed technique with a dinner napkin. . . . I gnash my teeth to think how I have misunderstood myself and my art" (quoted in Hicks 2005, 130).

59. Hicks 1991.

60. Ibid., 100.

61. On sex offender laws of the late 1930s, see Freedman 1987.

62. Hicks 2005, 105.

63. The election of Democratic governor Cuthbert Olson in 1938 initiated a period of prison reform in California and the commuting of sentences of several convicted "sex offenders." Cowell was released on parole in 1940 and married shortly thereafter. See Hicks 1991.

Chapter Five

1. Morantz 1977, 569.

2. Igo 2007, 222.

3. Ericksen 1999.

4. On the history of marriage education courses on American campuses, and Kinsey's place within that movement, see Bailey 1987.

5. See, for example, Pearl 1922. For an insightful account of Pearl and Reed's experiments, and of Pearl's life and work within the emerging field of population biology, see Kingsland 1985.

6. On the marginalization of Pearl within the emerging field of demography, see Ramsden 2002.

7. Jones 1997, 359.

8. Ibid., 72.
9. See Gebhard and Johnson 1998, 16.
10. Ibid., 73.
11. Latour 1990, 3. See particularly Latour and Woolgar 1985; Latour 1983, 1987, and 1993.
12. Kohler (2002, 91–93) notes that the gall wasp was ideally suited to naturalistic study but was resistant to being transported back to the laboratory and to becoming an experimental animal.
13. Jones 1997, 469.
14. Ericksen 1998, particularly 135–36.
15. For a cogent, if partial, summary of such criticisms, see Pomeroy 1972, 283–306.
16. Specifically, Cole Porter's song "Too Darn Hot" contained the line "according to the Kinsey report any average man you know" when it appeared in the Broadway musical *Kiss Me Kate* in 1948, but the "Kinsey report" was replaced by the "weather report" in the 1953 movie version by MGM. See Adams 2006.
17. Elsewhere I develop the point that the convincingness of graphs may be more likely to occur when scientists communicate across disciplinary boundaries in which graphing conventions differ, or across status boundaries, as with students or the public. See Hegarty and Walton 2012.
18. See also Shapin 2008.
19. Lewis Terman, Letter to Robert M. Yerkes, February 3, 1953. TP.
20. Pomeroy 1972, 203.
21. For an insightful criticism of the positionality of Kinsey and his interviewers, see Irvine 1990.
22. "The questions which you raise at the bottom of page 454 could perhaps be answered in terms of my approach, with this addition: nocturnal emissions are inversely correlated with other forms of outlet, especially heterosexual outlet. Since the amount of heterosexual experience is less among those who move up to a professional class, it is not too surprising to find the differential experience which you find inexplicable." Eli Ginzberg, Letter to Lewis M Terman, December 28, 1948. TP.
23. Latour 1987.
24. Christian 1971, 110.
25. Jones 1997, 434.
26. Kissack 2000. See also Minton 2002 on gay men's involvements with Kinsey's research.
27. Cochran, Mosteller, and Tukey 1954.
28. Lewis M. Terman, Letter to Robert M. Yerkes, March 9, 1948. TP.
29. Robert M. Yerkes, Letter to Lewis M. Terman, May 21, 1948. TP.
30. Igo 2007, 270.
31. On behaviorism, see particularly Lemov 2005; Mills 1998; Rutherford 2009; Smith 1986.
32. On Foucault's opposition to the more psychoanalytic "desire" in favor of "pleasure" as a term for understanding sex and sexual politics, see Foucault [1978] 1997.

33. See, for example, Thagard 2006.
34. This is not to say that behaviorists did not issue a response to Festinger's challenge. Building on B. F. Skinner's theory of "private events," social psychologist Daryl Bem's self-perception theory later forwarded the claim that the results of Festinger's experiments did not require one to posit the state of "cognitive dissonance" but could be predicted by a less elaborate theory that people simply inferred what their attitudes might be from observing their own behavior. The debate between dissonance and self-perception theories of attitude change preoccupied experimental social psychologists particularly in the late 1960s and early 1970s. For an influential review of this experimental work see Aronson 1969.
35. Lubek and Stam 1995.
36. See, for example, Kelman 1967; Sasson and Nelson 1969.
37. See Brunner 2001; Buchanan 1997; Lemov 2011; Wood et al. 2003.
38. On the rise and fall in faith in the Rorschach test's ability to detect the homosexual personality, see Hegarty 2003a.

Chapter Six

1. Jones 1997, 115–20.
2. Pomeroy 1972, 29.
3. Glenn Ramsey, Letter to Alfred C. Kinsey, December 12, 1941. KP. Earlier in 1941, Thurman Rice had attacked the legitimacy of Ramsey's project at a Chicago conference. See Glenn Ramsey, Letters to Alfred C. Kinsey, November 7, 1939; January 23, 1940; December 12, 1941. KP. On Ramsey's studies in Peoria see also Ramsey 1943.
4. A. H. Hobbs, Letter to Alfred C. Kinsey, October 13, 1947. KP.
5. On the persistently conflictual relationships between religious fundamentalism in the United States and the changes in sexual morality affected by Kinsey's research, see Cobb 2006; Herman 1997; Irvine 2002; Martin 2005.
6. For an example of criticism that appeals both to religious and secular sources of authority, including Terman's review, see Daniels 1954.
7. Sander Gilman (1996) critiques the failure to discount the romanticism of shtetl life by the informants and their parents who would have left the shtetls for Central European cities. Joyce Antler (2008) describes *Life is with People* as representing Jewish culture as timeless and Jewish mothers as endlessly self-sacrificing, eclipsing other accounts of American Jewish women as modern, flexible, and pragmatic kinds of people abroad in the social sciences of the day.
8. See Samelson 1978.
9. As Graham Richards (1997, 138–39) notes, Jewish names were common among psychologists who studied "race prejudice" in this era.
10. On modern prejudice, see McConahay 1983; Morrison and Morrison 2002; Swim et al. 1995.
11. In contrast, Kinsey's work aimed at limiting the "projections of the id" onto African Americans. Kinsey collected sexual histories from many African American interviewees but did not consider that he had collected a sufficient

number for inclusion in *SBHM*. Therein he emphasized caution in attributing differences to race that might be explicable in terms of the effects of class on sexual patterns, describing the histories of Whites and African Americans as "close if not identical" when class was taken into account. Where Kinsey described individual African American men, he tended to do so to exemplify atypical and valued sexual expression such as surprisingly frequent ejaculations, or surprisingly long-lived sexual activity into old age (*SBHM*, 217, 235–37, 393).

12. Like many intellectuals of his generation, Trilling was distrustful of popular culture forms that circulated more widely than in intellectual circles as useful grounds for cultural production. See Andrew Ross 1989.

13. I use the term "contrapunctual" here in the sense meant by Edward Said (2004, 24) to imply that Angus's writing "travels across temporal, cultural and ideological boundaries in unforeseen ways to emerge as part of a new ensemble *along with* later history and subsequent art."

14. On Angus's life and work, see Emilson 1991.

15. Ibid., 24.

16. Gilman (1996) offers an extensive bibliography of commentary on *Moses* in the half century since its publication. Scholars of recent decades have not avoided the text either. See particularly Derrida 1995; Said 2004; Yerushalmi 1993.

17. Freud's value judgment about this choice became bolder over successive writings. In an earlier version this passage read as "the alternative that was more significant *culturally*" (my emphasis).

18. See particularly the bibliography of writing on *Moses and Monotheism* in Gilman 1993.

19. Lewis M. Terman, Letter to Robert M. Yerkes, June 20, 1949. TP.

20. Crosby and Hastorf 2000, 146.

21. Quoted in Seagoe 1975, 226–27.

22. Storfer 1990; see also Seagoe 1975, 275.

23. As ever, Terman was reluctant to blame husbands for differences among wives' experiences of married sex: "This may mean either that strict religious training on the part of the husband is unfavorable to establishment of orgasms in the wife, or that men of the religious sort are more likely to mate with the type of woman who is inadequate" (Terman 1938, 395–96).

24. Lewis M. Terman, Letter to Robert M. Yerkes, June 20, 1949. TP.

25. See Blum 2002. Terman had grounds to be worried about how Israel/Harlow might be stereotyped because of his name. Terman's friend E. G. Boring often discussed the presence or absence of the "objectionable traits" of Jewish psychologists in his letters of recommendation. Terman had also asked Boring if the German social psychologist Kurt Lewin, Festinger's mentor, possessed such "objectionable traits" as those that suggested he had emigrated from Europe only recently, or had retained European Jewish traditions consistent with his understanding of assimilated, Americanized Jews as particularly smart. This exchange occurred in the context of Terman considering Lewis as a replacement for Walter Miles after Miles's departure for Yale. While Lewin did visit Stanford in 1932, Terman never hired Lewin, citing budget cuts as the reason for his decision (Winston 1996 and 1998).

Chapter Seven

1. Quoted in Jones 1997, 571.
2. See Bush 1945. See also the following speeches by outgoing ASA presidents in the postwar period: Lubin 1947; Thorp 1948; Walker 1945. On the growth of the membership of the ASA during the postwar period, see Stigler 1989.
3. Haraway, "The Biological Enterprise: Sex, Mind, and Profit from Human Engineering to Sociobiology" and "A Cyborg Manifesto: Science, Technology and Socialist-Feminism in the Late Twentieth Century," in *Simians, Cyborgs and Women* (1991, 43–68, 127–48).
4. Haraway (1991, 14) notes only that Yerkes, through the CRPS, funded Kinsey's studies.
5. See Ericksen 1998.
6. Gigerenzer et al. 1989; Hacking 1975 and 1990; Kruger, Gigerenzer, and Morgan 1987; Kruger, Daston, and Heiderberger 1987.
7. On the history of the US census, see Alterman 1969; Anderson 1988; Wright 1990.
8. Fitzpatrick 1957.
9. Walker 1890.
10. By the turn of the century, the question of the center of a population had become a technical one. See Hayford 1902.
11. For an insightful essay on the impact of the census on Turner, see Popper, Lang, and Popper 2000.
12. Fitzpatrick 1955.
13. Alterman 1969; Anderson 1988; Fitzgerald 1957; Wright 1990.
14. Haraway notes that "Michel Foucault's biopolitics is a flaccid premonition of cyborg politics, a very open field" and that "the cyborg is not subject to Foucault's biopolitics; the cyborg simulates politics, a much more potent field of operations" (1991, 150, 163).
15. For an account of the "splitting and inversion" of the relationship between objects and concepts, see Latour and Woolgar 1985, 176–79.
16. Ayers 1927; see also Glover 1926.
17. Davis 1937; King 1936; Mills 1935; Rice 1934.
18. See Anderson 1988, 159–90.
19. See also Hotelling 1927.
20. Igo 2006.
21. See, for example, Danziger 1997; Igo 2007; Rose 1996.
22. See, for example, Stephan 1939.
23. Hauser 1941.
24. Lowell Reed, Letter to Robert M. Yerkes, December 22, 1942. KP.
25. Ibid.
26. Lowell Reed, Letter to Alfred C. Kinsey, December 28, 1942. KP.
27. Alfred C. Kinsey, Letter to Lowell Reed, August 16, 1943. KP.
28. Lowell Reed, Letter to Alfred C. Kinsey, August 24, 1943. KP.
29. Alfred C. Kinsey, Letter to Robert M. Yerkes, February 1946. KP.
30. Robert M. Yerkes, Letter to Alfred C. Kinsey, September 25, 1943. KP.
31. Alfred C. Kinsey, Letter to Robert M. Yerkes, September 28, 1943. KP.

32. Kinsey discussed the calculation of statistics and the construction of the incidence curves in some detail for his lay readers (*SBHM*, 109–19).

33. In support of Kinsey's intuition, psychologists have continually applied Fisher's tests of significance with greater enthusiasm than understanding of their epistemological meaning. For a critique of psychologists' uses of significance testing, see Cohen 1994.

34. See, for example, Popper 1959. On the relationship between "what happened" and "what happens" in science, see Shapin 1996. For a recent and representative sample of psychological research on induction, see Feeney and Heit 2007.

35. Although Kinsey reported on an all-White sample, he made this point concrete with reference to race. "For instance, Negros constitute less than 10 per cent of the total population of the United States (U.S. Census, 1940); but a Negro sample that was only a tenth as large as a white sample would be much less adequate than the white sample. If one is to study Negros as a group, now should have as many Negro cases as white" (*SBHM*, 92).

36. Kinsey made this point concrete with reference to the incidence of masturbation and homosexuality among college men.

37. This criticism of the representativeness of Kinsey's sample was made directly and forcefully by Abraham H. Maslow and James M. Sakoda (1952). These authors found that volunteers for a Kinsey-like study tended to score higher on a measure of self-esteem than non-volunteers. Subsequent studies have repeatedly found that people who volunteer to participate in sex research are more liberal in their attitudes to sex and more varied in their sexual experiences. See Bogaert 1996; Morokoff 1986; Siegman 1956; Wiederman 1999.

38. Hyman and Sheatsley 1948; Goldstein and Pastore 1948.

39. Hastorf, Hilgard, and Sears 1988; McNemar 1980.

40. Rucci and Tweney 1980. During this period psychologists also began to split and invert the relationship between the "variables" that they measured in their experiments and the people who participated in them, imagining people to possess clusters of personality "variables" that preexisted the psychologists' measurement of them (Danziger and Dzinis 1997).

41. In the first review, McNemar (1940) critiqued the tendency of psychologists to construct samples based entirely on college students. As Kurt Danziger (1990) has noted, college students had become the "normative" human participants in psychological research in the interwar decades. Not only did the proportion of studies that relied on college student participants increase, but college students were also less likely to be described in terms of their institutional location than participants recruited through schools, hospitals, armies, or other institutions. The problem of overreliance on college students did not abate in response to McNemar's critique but continues to this day. Indeed, it is one of the characteristic aspects of American psychology that has been globalized. See Sears 1986 and Moghaddam and Lee 2006. In the second article, McNemar (1946) concluded that the army studies of the early 1940s had not advanced sampling beyond the practices used in the 1940 census.

42. Alfred C. Kinsey, Letter to Lewis M. Terman, September 27, 1946. KP.

43. Lewis M. Terman, Letter to Robert M. Yerkes, May 18, 1949. TP.
44. Robert M. Yerkes, Letter to Lewis M. Terman, October 14, 1948. YP.
45. Lewis M. Terman, Letter to Robert M. Yerkes, October 19, 1948. TP.
46. Robert M. Yerkes, Letter to George Corner, October 26, 1948. YP.
47. Robert M. Yerkes, Letter to Alfred C. Kinsey, December 6, 1948. YP.
48. Robert M. Yerkes, Letter to Lewis M. Terman, December 14, 1948. YP.
49. Lewis M. Terman, Letter to Robert M. Yerkes, December 21, 1949. TP.
50. George Corner, Letter to Robert M. Yerkes, December 1, 1948. YP.
51. W. A. Wallis, Letter to Lewis M. Terman, November 8, 1948. TP.
52. Merrill Flood, Letter to Alfred C. Kinsey, November 8, 1948. KP.
53. Alfred C. Kinsey, Letter to Merrill Flood, November 30, 1948. KP.
54. Lowell Reed, Letter to Alfred C. Kinsey, December 10, 1948. KP.
55. Alfred C. Kinsey, Letter to Lowell Reed, January 6, 1949. KP.
56. Lewis M. Terman, Letter to W. A. Wallis, January 20, 1949.
57. Lewis M. Terman, Letter to Albert Ellis, January 4, 1949. TP.
58. Albert Ellis, Letter to Lewis Terman, January 14, 1949. TP.
59. Lewis Terman, Letter to Albert Ellis, January 25, 1949. TP.
60. Hegarty 2007.
61. Norman Fenton, Letter to Lewis M. Terman, January 11, 1948; Lewis M. Terman, Letter to Norman Fenton, January 25, 1948. TP.
62. See, for example, Burgess 1947.
63. Anon. 1948.
64. Jones 1997, 639.
65. Alfred C. Kinsey, Letter to Samuel Wilks, August 28, 1950. KP.
66. William Cochran, Letter to Alfred C. Kinsey, March 28, 1951. KP.
67. "We are also very much disturbed by the fact that I have known, in the course of my years of research, a great many people who have been diverted from studying their subject matter, and become completely immersed in the problem of method. The history of psychological studies is an unfortunate demonstration of the fact that people can become much more interested in methodological study than in the subject matter which they set out to investigate." Alfred C. Kinsey, Letter to William Cochran, November 6, 1951. KP.
68. See Ericksen 1998 and Jones 1997.
69. See folder on News Articles Published in Response to the report of the ASA published in June 1952. KI.
70. *New York Post*, Sunday, June 22, 1952.

Chapter Eight

1. Indeed, as I argued (1997) some time ago, Butler's scheme also illuminates attempts within neuroscience to ontologize the brains of gay men as "feminine."
2. On the relationship between Terman and Miles's M-F test and the MMPI, see Buchanan 1994.
3. See Constantinople (1973) for a psychometric critique and Lewin (1984) for a historical critique.
4. Psychologist Richard Lippa (2005) located his studies of "gender-diagnosticity" firmly within the tradition originating with *Sex and Personal-*

ity. For a discussion of the legacy of Terman and Miles thinking in sexology more generally, see Sandfort 2005. For a recent debate about the use of gender-typing measures to discern sexual orientation, see Hegarty 2009a; Rieger, Linsenmeier and Bailey 2009.

5. Karl Bryant's history (2006, 2008) of the formation of the diagnosis of "Gender Identity Disorder in Childhood" shows how little gay/lesbian spokespersons were involved in the diagnostic debates, with material effects for the shape of the diagnostic category. By cisgenderism, I mean the ideology that privileges the alignment of gender identity with external gender assignation; see Ansara and Hegarty 2012.

6. For a longer argument about the figuration of children as the target of queer theory critique, see Hegarty 2009b.

7. For a quantitative approach to childhood gender identity that neither relies upon the gender typing of activities nor conflates atypicality with pathology, see Egan and Perry 2001.

8. See, for example, Azurmendi et al. 2006.

9. See, for example, Auyeung et al. 2009 and Manning et al. 2001. For an overview of the "extreme male brain" theory of autism, see Baron-Cohen 2002.

10. See Hegarty 2011.

11. Kendall 1999. Kendall argues that the rise of microcomputers is responsible for the shift in nerd narratives over this period. However, I think that the "queerness" of high male intellect continues to haunt even the most celebratory narratives that bring together the Northern Californian frontier of Silicon Valley (brought about in no small measure by the gifted child Fred Terman), capitalist success, and individual willingness to work endlessly on abstract problems. Robert Cringely's history of the rise of microcomputing industry, *Accidental Empires* (1996), is quite deceptively subtitled "How the boys of Silicon Valley make their millions, battle foreign competition and still can't get a date." The book's title is a false premise, and it says very little about the romantic lives of the "boys." And yet, it seems necessary to flag the possibility of failed development, and particularly of blighted heterosexual development, as a consequence of nerdiness.

12. Quoted in Byers 1996, 420.

13. Ibid., 433.

14. See Ericksen 1999; Gebhard, Johnson, and Kinsey 1979; Irvine 1990.

15. On the development of the "sex survey" genre after Kinsey, see particularly Ericksen 1999.

16. Pruitt 2002.

17. Because it has been well discussed elsewhere I have chosen not to belabor the additional point that Kinsey's study also led many gay and lesbian people to imagine a community of like-minded people, and also led to a panic about homosexuals in government employment in the decade after its publication. The politics of small and large numbers are complex, and more invested with emotion and self-relevance than ahistorical debates suggest. See Cuordileone 2000; D'Emilio 1983; Reumann 2005; Terry 1999, 329–52.

18. Logan 2006, 86.

19. Sleigh 2006, 281.

Works Cited

Abelove, Henry. 1993. "Freud, Male Homosexuality, and the Americans."
In *The Lesbian and Gay Studies Reader*, ed. Henry Abelove, Michele Barale, and David M. Halperin, 381–96. New York: Routledge.

Aberle, Sophie D., and George W. Corner. 1953. *Twenty Five Years of Sex
Research: History of the National Research Council Committee for Research
in Problems of Sex, 1922–1947*. Philadelphia: W. B. Saunders & Company.

Adams, Naomi. 2006. "Kiss Me Kate: A New View of Women and Sex."
Lesbian & Gay Psychology Review 7:276–80.

Allport, Gordon. 1954. *The Nature of Prejudice*. Garden City, N.J.:
Doubleday.

Alterman, Hyman. 1969. *Counting People: The Census in History*. New York:
Harcourt Brace & World, Inc.

Amigot, Patricia, and Margot Pujal. 2009. "On Power, Freedom, and
Gender: A Fruitful Tension between Foucault and Feminism." *Theory &
Psychology* 19:646–69.

Anderson, Margo. 1988. *The American Census: A Social History*. New Haven:
Yale University Press.

Angus, Samuel. 1925. *The Mystery Religions and Christianity: A Study in the
Religious Background of Early Christianity*. New York: Charles Scribner's
Sons. Rpt., Philadelphia: W. B. Saunders & Company, 1947.

Anon. 1948. "Committees and Association Representatives for 1948." *The
American Statistician* 2:21.

Anon. 1986. "Psychological Professional Gold Medal Award: E. Lowell
Kelly." *American Psychologist* 41:411–12.

Ansara, Y. Gavriel, and Peter Hegarty. 2012. "Cisgenderism in Psychology:

Pathologizing and Misgendering Children from 1999 to 2008." *Psychology & Sexuality* 3:137–60.

Antler, Joyce. 2008. *You Never Call! You Never Write! A History of the Jewish Mother*. New York: Oxford University Press.

Arnett, Jeffrey Jenson. 2006. "G. Stanley Hall's Adolescence: Brilliance and Nonsense." *History of Psychology* 9:186–97.

Aronson, Elliot. 1969. "The Theory of Cognitive Dissonance: A Current Perspective." *Advances in Experimental Social Psychology* 4:2–34.

———. 2007. "Elliot Aronson." In *A History of Psychology in Autobiography*, vol. 9, ed. Gardner Lindzey. Washington, D.C.: American Psychological Association.

Aronson, Elliot, and Judson Mills, 1959. "The Effect of Severity of Initiation on Liking for a Group." *Journal of Abnormal and Social Psychology* 59:177–81.

Auyeung, Bonnie, Simon Baron-Cohen, Emma Ashwin, Rebecca Knickmeyer, Kevin Taylor, and Gerald Hackett. 2009. "Fetal Testosterone and Autistic Traits." *British Journal of Psychology* 100:1–22.

Ayers, Leonard P. 1927. "The Dilemma of the New Statistician." *Journal of the American Statistical Association* 22:1–8.

Azurmendi, Aitziber, Francisco Braza, Ainhoa García, Paloma Braza, José M. Muñoz, and José Sánchez-Martín. 2006. "Aggression, Dominance, and Affiliation: Their Relationships with Androgen Levels and Intelligence in 5-year-old Children." *Hormones and Behavior* 50:132–40.

Bailey, Beth L. 1987. "Scientific Truth . . . And Love: The Marriage Education Movement in the United States." *Journal of Social History* 20:711–732.

Bailey, J. Michael, and Kenneth J. Zucker. 1995. "Childhood Sex-Typed Behavior and Sexual Orientation: A Conceptual Analysis and Quantitative Review." *Developmental Psychology* 31:43–55.

Bakan, David. 1958. *Sigmund Freud and the Jewish Mystical Tradition*. Princeton: Von Nostrand.

Baron-Cohen, Simon. 2002. "The Extreme Male Brain Theory of Autism." *Trends in Cognitive Science* 6:248–54.

Bates, Barbara. 1992. *A Social History of Tuberculosis, 1876–1939*. Philadelphia: University of Pennsylvania Press.

Battersby, Christine. 1989. *Gender and Genius: Toward a Feminist Aesthetics*. Bloomington: Indiana University Press.

Bayer, Ronald. 1981. *Homosexuality and American Psychiatry: The Politics of Diagnosis*. New York: Basic Books.

Becker, George. 1978. *The Mad Genius Controversy: A Study in the Sociology of Deviance*. Beverly Hills: Sage.

Bem, Daryl J. 1996. "Exotic Becomes Erotic: A Developmental Theory of Sexual Orientation." *Psychological Review* 103:320–35.

Bem, Sandra Lipsitz. 2001. *An Unconventional Family*. New Haven: Yale University Press.

Bertolini, Vincent J. 1996. "Fireside Chastity: The Erotics of Sentimental Bachelorhood in the 1850s." *American Literature* 68:707–37.

Bérubé, Allan. 1990. *Coming Out Under Fire: The History of Gay Men and Women in World War II*. New York: Free Press.

Benjamin, Ludy A., and Stephanie A. Shields. 1990. "Leta Stetter Hollingworth (1886–1939)." In *Women in Psychology: A Bio-Biographic Sourcebook*, ed. Agnes N. O'Connell and Nancy F. Russo, 173–83. Westport, Conn,: Greenwood Press.

Benson, Ciaran. 1987. "Ireland's Low IQ: A Critique of the Myth." *Irish Journal of Psychology* 8:61–70.

Billig, Michael. 1987. *Arguing and Thinking: A Rhetorical Approach to Social Psychology*. Cambridge: Cambridge University Press.

———. 1995. *Banal Nationalism*. London: Sage.

Binet, Alfred, and Theodore Simon. 1905. "Méthodes nouvelle pour le diagnostic de niveau intellectuel des anormaux." *L'Année Psychologique* 11:191–244.

Block, Ned J., and Gerald Dworkin. 1976. *The I.Q. Controversy: Critical Readings*. New York: Random House.

Blum, Deborah. 2002. *Love at Goon Park: Harry Harlow and the Science of Affection*. Cambridge, Mass.: Perseus.

Bogaert, Anthony F. 1996. "Volunteer Bias in Human Sexuality Research: Evidence for Both Sexuality and Personality Differences in Males." *Archives of Sexual Behavior* 25:125–40.

Bogue, Donald J. 1974. "Introduction." In *The Basic Writings of Ernest W. Burgess*, ed. Donald J. Bogue, ix–xxiv. Chicago: University and Family Study Center.

Bond, Horace Mann. 1924. "Intelligence Tests and Propaganda." *The Crisis* 28 (June): 61–64.

———. 1926. "An Investigation of the Non-Intellectual Traits of a Group of Negro Adults." *Journal of Abnormal and Social Psychology* 21:267–76.

Boring, E. G. June 1923. "Intelligence as the Tests Test It." *New Republic*, 6 June, 35–37.

———. 1929. *A History of Experimental Psychology*. New York: Appleton-Century-Crofts.

Boyle, T. C. 2004. *The Inner Circle*. New York: Viking.

Bramel, Dana. 1963. "Selection of a Target for Defensive Projection." *Journal of Abnormal and Social Psychology* 66:318–24.

Brandt, Allan M. 1985. *No Magic Bullet: A Social History of Venereal Disease in the United States since 1880*. New York: Oxford University Press.

Brown, Eric C., ed. 2006. *Insect Poetics*. Minneapolis: University of Minnesota Press.

Brown, Eric C. 2006. "Introduction: Reading the Insect." In *Insect Poetics*, ed. Brown, ix–xxiii.

Brown, JoAnne. 1992. *The Definition of a Profession: The Authority of Metaphor in the History of Intelligence Testing*. Princeton: Princeton University Press.

Brunner, José. "'Oh Those Crazy Cards Again': A History of the Debate on the Nazi Rorschachs, 1946–2001." *Political Psychology* 22:223–61.

Bryant, Karl. 2006. "Making Gender Identity Disorder of Childhood: Historical Lessons for Contemporary Debates." *Sexuality Research and Social Policy* 3:23–39.

———. 2008. "In Defense of Gay Children? 'Progay' Homophobia and the Production of Homonormativity." *Sexualities* 11:455–75.

Buchanan, Roderick D. 1994. "The Development of the Minnesota Multiphasic Personality Inventory." *Journal of the History of the Behavioral Sciences* 30:148–61.

———. 1997. "Ink Blots or Profile Plots: The Rorschach versus the MMPI as the Right Tool for a Science-Based Profession." *Science, Technology & Human Values* 22:168–206.

Budge, G. Scott, and Bernard Katz. 1995. "Constructing Psychological Knowledge: Reflections on Science, Scientists and Epistemology in the APA Publication Manual." *Theory & Psychology* 5:217–31.

Buenker, John D., and Edward R. Kantowitz. 1988. *Historical Dictionary of the Progressive Era, 1890–1920*. New York: Greenwood Press.

Bullough, Vern L. 1987. "Technology for the Prevention of 'Les Maladie produites par la masturbation.'" *Technology and Culture* 28:828–32.

———. 2004. "Sex Will Never Be the Same Again: The Contributions of Alfred C. Kinsey." *Archives of Sexual Behavior* 33:277–86.

Burgess, Ernest W., and Leonard S. Cottrell Jr. 1939. *Predicting Success or Failure in Marriage*. New York: Prentice Hall.

Burgess, R. W. 1947. "Do We Need a 'Bureau of Standards' for Statistics?" *Journal of Marketing* 11:281–82.

Burman, Erica, and Ian Parker, I. 1993. *Discourse Analytic Research: Repertoires and Readings of Texts in Action*. London: Routledge.

Bush, Vannevar. 1945. *Science, the Endless Frontier: A Report to the President on a Program for Postwar Scientific Research*. Washington, D.C.: Government Printing Office.

Buss, Allan R. 1978. "The Structure of Psychological Revolutions." *Journal of the History of the Behavioral Sciences* 14:57–64.

Butler, Judith. 1990. *Gender Trouble: Feminism and the Subversion of Identity*. New York: Routledge.

———. 1993. *Bodies That Matter: On the Discursive Limits of 'Sex.'* New York: Routledge.

Byers, Thomas B. 1996. "History Re-membered: Forrest Gump, Postfeminist Masculinity, and the Burial of the Counterculture." *Modern Fiction Studies* 42:419–44.

Capshew, James H., Matthew H. Adamson, Patricia A. Buchanan, Narisara Murray, and Naoko Wake. 2003. "Kinsey's Biographers: A Historical Reconnaissance." *Journal of the History of Sexuality* 12:465–86.

Carson, John. 1993. "Army Alpha, Army Brass and the Search for Army Intelligence." *Isis* 84:278–309.

———. 2007. *The Measure of Merit: Talents, Intelligence and Inequality in the French and American Republics, 1750–1940*. Princeton: Princeton University Press.

Cattell, James McKeen. 1890. "Mental Tests and Measurement." *Mind* 15:373–81.

———. 1906. *American Men of Science*. New York: Bowker.

Chapman, Paul Davis. 1988. *Schools as Sorters: Lewis M. Terman, Applied Psychology, and the Intelligence Testing Movement, 1890–1930*. New York: New York University Press.

Cherry, Frances. 1995. *The Stubborn Particulars of Social Psychology*. London: Routledge.

Chesler, Ellen. 1992. *Woman of Valor: Margaret Sanger and the Birth Control Movement in America*. New York, Simon & Schuster.

Chiang, Howard Hsueh-Ho. 2008. "Effecting Science, Affecting Medicine: Homosexuality, the Kinsey Reports, and the Contested Boundaries of Psychopathology in the United States, 1948–1965." *Journal of the History of the Behavioral Sciences* 44:300–18.

Christenson, Cornelia V. 1971. *Kinsey: A Biography*. Bloomington: Indiana University Press.

Clarke, Adele E. 1990. "Controversy and the Development of the Reproductive Sciences." *Social Problems* 37:18–37.

———. 1998. *Disciplining Reproduction: Modernity, American Life Sciences, and the Problems of Sex*. Berkeley and Los Angeles: University of California Press.

Cobb, Michael. 2006. *God Hates Fags: The Rhetorics of Religious Violence*. New York: New York University Press.

Cochran, William G., Frederick Mosteller, and John W. Tukey. 1954. *Statistical Problems of the Kinsey Report on Sexual Behavior in the Human Male*. Washington, D.C.: American Statistical Association.

Cohen, Jacob. 1994. "The Earth Is Round, p <.05." *American Psychologist* 49:997–1003.

Condon, Bill, director. 2004. *Kinsey*. Fox Searchlight Pictures.

Constantinople, Anne. 1973. "Masculinity-Femininity: An Exception to a Famous Dictum?" *Psychological Bulletin* 80:389–407.

Cowell, Henry. 1926. "The Process of Musical Creation." *The American Journal of Psychology* 37:233–36.

Cox, Catharine Morris. 1926. *Genetic Studies of Genius*, vol. 2: *The Early Mental Traits of Three Hundred Geniuses*. Stanford: Stanford University Press.

Cravens, Hamilton. 1992. "A Scientific Project Locked in Time: The Terman Genetic Studies of Genius, 1920s–1950s." *American Psychologist* 47:183–89.

———. 1993. *Before Head Start: The Iowa Station & America's Children*. Chapel Hill: The University of North Carolina Press.

———. 2006. "The Historical Context of G. Stanley Hall's *Adolescence* (1904)." *History of Psychology* 9:172–85.

Cringely, Robert X. 1996. *Accidental Empires*. 2d ed. London: Penguin.

Crosby, Jennifer Randall, and Albert Hastorf. 2000. "Lewis Terman: Scientist of Mental Measurement and Product of His Time." In *Portraits of Pioneers in Psychology*, ed. Gregory A. Kimble and Michael Wertheimer, 137–54. Washington, D.C.: American Psychological Association.

Cuordileone, K. A. 2000. "'Politics in an Age of Anxiety': Cold War Political Culture and the Crisis in American Masculinity, 1949–1960." *Journal of American History* 87:515–45.

Curtain, Tyler. 1997. "The 'Sinister Fruitiness' of Machines: *Neuromancer*, Internet Sexuality, and the Turing Test." In *Novel Gazing: Queer Readings in Fiction*, ed. Eve Kosofsky Sedgwick, 128–48. Durham, N.C.: Duke University Press.

Daniels, E. J. 1954. *I Accuse Kinsey!* Orlando: Christ for the World Publishers.

Danziger, Kurt. 1987. "Statistical Method and the Historical Development of Research Practices in American Psychology." *The Probabilistic Revolution*, vol. 2: *Ideas in the Sciences*, ed. Lorenz Kruger, Gerd Gigerenzer, and Mary S. Morgan, 35–47. Cambridge, Mass.: MIT Press.

———. 1990. *Constructing the Subject: Historical Origins of Psychological Research*. Cambridge: Cambridge University Press.

———. 1991. "Guest Editor's Introduction." *History of the Human Sciences* 4:327–33.

———. 1997. "Varieties of Social Constructionism." *Theory & Psychology* 7:399–416.

———. *Naming the Mind: How Psychology Found Its Language*. London: Sage.

Danziger, Kurt, and Katalin Dzinis. 1997. "How Psychology Got Its Variables." *Canadian Psychology* 38:43–48.

Darwin, Charles. 1859. *On the Origin of Species*. London: John Murray.

Daston, Lorraine J. 1987. "Rational Individuals Versus Laws of Society: From Probability to Statistics." In *The Probabilistic Revolution*, vol. 1: *Ideas in History*, ed. Lorenz Kruger, Lorraine J. Daston, and Michael Heiderberger, 295–304. Cambridge, Mass.: MIT Press, 1987.

Davis, Joseph S. 1937. "Statistics and Social Engineering." *Journal of the American Statistical Association* 32:1–7.

Davis, Katherine. 1929. *Factors in the Sex Lives of Twenty-Two Hundred Women*. New York: Harper & Brothers.

D'Emilio, John. 1993. "Capitalism and Gay Identity." In *The Lesbian and Gay Studies Reader*, ed. Henry Abelove, Michele Barale, and David M. Halperin, 467–78. New York: Routledge, 1993.

Derrida, Jacques. 1995. *Archive Fever: A Freudian Impression*. Translated by Eric Prenowitz. Chicago: University of Chicago Press.

Dewsbury, Donald A. 2006. *Monkey Farm: A History of the Yerkes Laboratories of Primate Biology, Orange Park, Florida, 1930–1965*. Cranbury, N.J.: Bucknell University Press.

Diehl, Leslie A. 1986. "The Paradox of G. Stanley Hall: Foe of Coeducation and Educator of Women." *American Psychologist* 41:868–78.

Dorfman, Joseph. 1984. *Thorstein Veblen and His America*. New York: Viking.

Douglas, M. 1966. *Purity and Danger: An Analysis of Concepts of Pollution and Taboo*. London: Routledge and Kegan Paul.

Driver-Linn, Erin. 2003. "Where is Psychology Going? Structural Fault Lines Revealed by Psychologists' Use of Kuhn." *American Psychologist* 58:269–78.

Duggan, Lisa. 2000. *Sapphic Slashers: Sex, Violence and American Modernity.* Durham, N.C.: Duke University Press.

Eby, Clare Virginia. 1999. "The Two Mrs Veblens, Among Others." *International Journal of Politics, Culture, and Society* 13:353–61.

Egan, Susan K., and David Perry. 2001. "Gender Identity: A Multidimensional Analysis with Implications for Psychosocial Adjustment." *Developmental Psychology* 37:451–63.

Ehrenreich, Barbara, and Deirdre English. 1979. *For Her Own Good: 150 Years of Experts' Advice to Women.* New York: Anchor Books.

Elfenbein, Andrew. 1998. *Romantic Genius: The Prehistory of a Homosexual Role.* New York: Columbia University Press.

Ellis, Albert. 1948. "Review of Kinsey, A. C., Pomeroy, W. B., and Martin, C. E. *Sexual Behavior in the Human Male.* Philadelphia: Saunders, 1948. Pp. 804." *Journal of General Psychology* 39:299–326.

Ellis, Havelock. 1934. *Man and Woman.* 8th ed. rev. London: William Heinemann.

———. 1936. *Studies in the Psychology of Sex*, vol. 2: *Sexual Inversion.* New York: Random House.

Emilson, Susan. 1991. *A Whiff of Heresy: Samuel Angus and the Presbyterian Church in New South Wales.* Sydney: University of New South Wales Press.

Ericksen, Julia A. 1998. "With Enough Cases, Why Do You Need Statistics? Revisiting Kinsey's Methodology." *Journal of Sex Research* 35:132–40.

———. 1999. *Kiss and Tell: Surveying Sex in the Twentieth Century.* Cambridge, Mass.: Harvard University Press.

Faderman, Lillian. 1991. *Odd Girls and Twilight Lovers: A History of Lesbian Life in Twentieth Century America.* New York: Columbia University Press.

Fairclough, Norman. 2003. "'Political Correctness': The Politics of Culture and Language." *Discourse & Society* 14:17–28.

Fancher, Raymond E. 1985. *The Intelligence Men: Makers of the IQ Controversy.* New York: W. W. Norton.

———. 1998. "Alfred Binet, General Psychologist." In *Portraits of Pioneers in Psychology*, vol. 3, ed. Gregory A. Kimble and Michael Wertheimer, 67–83. Washington, D.C.: American Psychological Association.

Faragher, John Mark. 1994. *Rereading Frederick Jackson Turner: "The Significance of the Frontier in American History" and Other Essays.* New Haven: Yale University Press.

Feeney, Aidan, and Evan Heit. 2007. *Inductive Reasoning: Experimental, Developmental and Computational Approaches*. Cambridge: Cambridge University Press.

Festinger, Leon. 1957. *A Theory of Cognitive Dissonance*. Stanford: Stanford University Press.

Festinger, Leon, and James M. Carlsmith. 1959. "Cognitive Consequences of Forced Compliance." *Journal of Abnormal and Social Psychology* 58:203–10.

Festinger, Leon, Henry W. Riecken, and Stanley Schachter. 1956. *When Prophecy Fails: A Social and Psychological Study of a Modern Group that Predicted the Destruction of the World*. New York: Harper-Torchwood.

Fisher, Ronald A. 1925. *Statistical Methods for Research Workers*. Edinburgh: Oliver & Boyd.

Fiske, Donald W., James J. Conley, and Lewis R. L. Goldberg. 1987. "E. Lowell Kelly (1905–1986)." *American Psychologist* 42:511–12.

Fitzpatrick, J. Paul. 1955. "The Early Teaching of Statistics in American Colleges and Universities." *The American Statistician* 9:12–18.

———. 1957. "Statistical Societies in the United States in the Nineteenth Century." *The American Statistician* 11:13–21.

Foschi, Renato, and Elisabetta Cicciola. "Politics and Naturalism in the 20th Century Psychology of Alfred Binet." *History of Psychology* 9:267–89.

Foucault, Michel. 1971. *The Order of Things: An Archaeology of the Human Sciences*. New York, Vintage.

———. 1977a. *Language, Counter-Memory, Practice: Selected Essays and Interviews*. Translated by Donald F. Bouchard and Sherry Simon. Ithaca, N.Y.: Cornell University Press.

———. 1977b. *Discipline and Punish: The Birth of the Prison*. Translated by Alan Sheridan. New York: Random House.

———. 1978. *The History of Sexuality, Volume 1: An Introduction*. Translated by Robert Hurley. New York: Vintage.

———. [1978] 1997. "Questions for Michel Foucault." Interview with Jean de Bitoux and Mattias Duyves. In *We Are Everywhere: A Historical Sourcebook of Gay and Lesbian Politics*, ed. Mark Blasius and Shane Phelan, 454–59. New York: Routledge.

———. 2003. *Abnormal: Lectures at the College De France, 1974–1975*. London: Verso.

Freedman, Estelle. 1987. "'Uncontrolled Desires': The Response to the Sexual Psychopath, 1920–1960." *Journal of American History* 74:83–106.

Freedman, Estelle, and John D'Emilio. 1992. *Intimate Matters: A History of Sexuality in America*. New York: Harper & Row.

Freud, Sigmund. 1905. "Three Essays on the Theory of Sexuality." In *The Standard Edition of the Complete Psychological Works of Sigmund Freud* (hereafter *SE*), trans. James Strachey, 7:125–245. London: Hogarth Press.

———. 1908. "Civilized Sexual Morality and Modern Nervous Illness." *SE* 9:181–204. London: Hogarth Press.

———. 1910. "Leonardo Da Vinci." *SE* 11:59–137. London: Hogarth Press.

———. 1925. "Some Psychical Consequences of the Anatomical Distinction between the Sexes." *SE* 19:243–58. London: Hogarth Press.

———. 1930. "Civilization and its Discontents." *SE* 21:59–148. London: Hogarth Press.

Freud, S. 1939. "Moses and Monotheism." *SE* 23:3–137. London: Hogarth Press.

Friedman, Howard S., Margaret L. Kern, and Chandra A. Reynolds. 2010. "Personality and Health, Subjective Well-being, and Longevity." *Journal of Personality* 78:179–216.

Furumoto, Lauren, and Elizabeth Scarborough. 1986. "The First American Women Psychologists." *American Psychologist* 41:35–42.

Galton, Francis. 1869. *Hereditary Genius: An Inquiry Into Its Laws and Consequences*. London: Macmillan & Co.

———. 1874. *English Men of Science: Their Nature and Nurture*. London: Macmillan & Co.

Gathorne-Hardy, Jonathan. 1998. *Sex the Measure of All Things: A Life of Alfred C. Kinsey*. London: Chatto & Windus.

Gebhard, Paul, Alan B. Johnson, and Alfred C. Kinsey. 1979. *The Kinsey Data: Marginal Tabulations of the 1938–1963 Interviews Conducted by the Institute for Sex Research*. Philadelphia: W. B. Saunders & Co.

Gergen, Kenneth J. 1985. "The Social Constructionist Movement in Modern Psychology." *American Psychologist* 40:266–75.

Gibson, Margaret. 1998. "The Masculine Degenerate: American Doctors' Portrayals of the Lesbian Intellect, 1880–1949." *Journal of Women's History* 9:78–102.

Gigerenzer, Gerd, Zeno Swijtink, Theodore Porter, Lorraine Daston, John Beatty, and Lorenz Kruger. 1989. *The Empire of Chance: How Probability Changed Science and Everyday Life*. Cambridge: Cambridge University Press.

Gilman, Sander. 1993. *Freud, Race, and Gender*. Princeton: Princeton University Press.

———. 1996. *Smart Jews: The Construction of the Image of Jewish Superior Intelligence*. Lincoln: University of Nebraska Press.

Ginzberg, Eli. 1948. "Sex and Class Behavior." In *About the Kinsey Report: Observations of 11 Experts on "Sexual Behavior in the Human Male,"* ed. Donald Porter Geddes and Enid Curie, 131–45. New York: Signet.

Glover, James W. 1926. "Requirements for Statisticians and Their Training: Statistical Teaching in American Colleges and Universities." *Journal of the American Statistical Association* 22:419–46.

Goddard, Henry H. 1910. "Four Hundred Feeble-Minded Children Classified by the Binet Method." *Pedagogical Seminary* 17:387–97.

———. 1923. *Feeble-Mindedness: Its Causes and Consequences*. New York: The Macmillan Company.

Goldstein, Jacob, and Nicholas Pastore. 1948. "Sexual Behavior of the American Male: A Special Review of the Kinsey Report." *The Journal of Psychology* 26:347–62.

Goodman, Barak, and John Maggio, directors. 2005. *American Experience: Kinsey*. Public Broadcasting Corporation.

Goodwin, James C. 1987. "In Hall's Shadow: Edmund Clark Sanford (1859–1924)." *Journal of the History of the Behavioral Sciences* 23:153–68.

Gordon, Linda. 1977. *Woman's Body, Woman's Right: Birth Control in America*. Middlesex, U.K.: Penguin.

Gottman, John M., and Clifford I. Notarius. 2002. "Marital Research in the 20th Century and a Research Agenda for the 21st Century. *Family Process* 41:159–97.

Gould, Stephen Jay. 1981. *The Mismeasure of Man*. New York: W. W. Norton & Company.

———. 1988. "Of Wasps and Wasps." In *The Flamingo's Smile*, 155–66. New York: Penguin.

Guthrie, Robert V. 1976. *Even the Rat was White: A Historical View of Psychology*. New York: Harper & Row.

———. 2000. "Sumner, Francis Cecil." In *Encyclopedia of Psychology*, vol. 7, ed. Alan E. Kazdin. Washington, D.C.: American Psychological Association.

Hacking, Ian. 1975. *The Emergence of Probability*. Cambridge: Cambridge University Press.

———. 1981. "How Should We Do a History of Statistics?" *Ideology & Consciousness* 8:16–26.

———. 1990. *The Taming of Chance*. Cambridge: Cambridge University Press.

Hall, Granville Stanley. 1885. "The New Psychology." *Andover Review* 3: 120–35, 239–48.

———. 1904. *Adolescence: Its Psychology and Its Relation to Physiology, Anthropology, Sociology, Sex, Crime, Religion and Education.* 2 vols. New York: Appleton.

———. 1923. *Life and Confessions of a Psychologist.* New York: Appleton.

Halperin, David M. 1995. *Saint Foucault: Toward a Gay Hagiography.* Oxford: Oxford University Press.

———. 2003. "The Normalization of Queer Theory." *Journal of Homosexuality* 45:339–43.

Halpern, Carolyn J. T., J. Richard Udry, Chirayath Suchindran, and Benjamin Campbell. 2000. "Adolescent Males' Willingness to Report Masturbation." *The Journal of Sex Research* 37:327–32.

Hamilton, George V. 1929. *A Research in Marriage.* New York: Albert and Charles Boni, Inc.

Haraway, Donna J. 1989. *Primate Visions: Gender, Race, and Nature in the Making of the Modern World.* New York: Routledge.

———. 1991. *Simians, Cyborgs and Women: The Reinvention of Nature.* New York: Routledge.

———. 1997. *Modest_Witness@Second Millenium.Femaleman©_Meets_ Oncomouse™.* New York: Routledge.

Harding, Sandra. 1987. *The Science Question in Feminism.* Ithaca, N.Y.: Cornell University Press.

Harper, Robert S. 1949. "Tables of American Doctorates in Psychology." *American Journal of Psychology* 62:579–87.

Hartsock, Nancy. 1990. "Foucault on Power: A Theory for Women?" In *Feminism/Postmodernism*, ed. Linda Nicholson, 157–75. New York: Routledge.

Hastorf, Albert H., Ernest R. Hilgard, and Robert R. Sears. 1988. "Obituary: Quinn McNemar (1900–1986)." *American Psychologist.* 43:196–97.

Hauser, Philip M. 1941. "The Use of Sampling in the Census." *Journal of the American Statistical Association* 36:369–75.

Hayford, J. F. 1902. "What is the Center of an Area or the Center of a Population?" *Publications of the American Statistical Association* 8:47–58.

Hegarty, Peter. 1997. "Materializing the Hypothalamus: A Performative Account of the 'Gay Brain.'" *Feminism & Psychology* 7:355–72.

———. 2001. "'Real Science,' Deception Experiments and the Gender of My Lab Coat: Toward a New Laboratory Manual for Lesbian and Gay Psychology." *International Journal of Critical Psychology* 1:91–108.

———. 2003a. "Homosexual Signs and Heterosexual Silences: Rorschach Studies of Male Homosexuality from 1921 to 1967." *Journal of the History of Sexuality* 12:400–23.

———. 2003b. "'More Feminine than 999 Men out of 1,000': Measuring Sex Roles and Gender Nonconformity in Psychology." In *Gender Nonconformity, Race, and Sexuality: Charting the Connections*, ed. Toni Lester, 62–83. Madison: University of Wisconsin Press.

———. 2005. "Harry Stack Sullivan and His Chums: Archive Fever in American Psychiatry?" *History of the Human Sciences* 18:35–53.

———. 2007. "Getting Dirty: Psychology's History of Power." *History of Psychology* 10:75–91.

———. 2007. "From Genius Inverts to Gendered Intelligence: Lewis Terman and the Power of the Norm." *History of Psychology* 10:132–55.

———. 2009a. "Toward an LGBT-Informed paradigm for Children who Break Gender Norms: Comment on Drummond et al. (2008) and Rieger et al. (2008)." *Developmental Psychology* 45:895–900.

———. 2009b. "Queerying Lesbian and Gay Psychology's Coming of Age: Was the Past Just Kid Stuff?" In *The Ashgate Research Companion to Queer Theory*, ed. Noreen Giffney and Michael O'Rourke, 311–28. Aldershot: Ashgate,.

———. 2011. "Sexuality, Normality, Intelligence: What is Queer Theory Up Against?" *Psychology & Sexuality* 2.1:45–57.

———. 2012. "Getting Miles Away from Terman: Did the CRPS Fund Catharine Cox Miles' Unsilenced Psychology of Sex?" *History of Psychology* 15:201–8.

Hegarty, Peter, and Felicia Pratto. 2001. "The Effects of Social Category Norms and Stereotypes on Explanations for Intergroup Differences." *Journal of Personality and Social Psychology* 80:723–735.

Hegarty, Peter, and Zoe Walton. 2012. "The Consequences of Predicting Scientific Impact in Psychology Using Journal Impact Factors." *Perspectives on Psychological Science* 7.1:72–78.

Henriques, Julian, Wendy Hollway, Cathy Urwin, Couze Venn, and Valerie Walkerdine. 1984. *Changing the Subject: Psychology, Social Regulation and Subjectivity*. London: Methuen & Co.

Henry, George. 1941. *Sex Variants: A Study in Homosexual Patterns*. 2 vols. New York: Paul B. Hoeber.

Herman, Didi. 1996. *The Antigay Agenda: Orthodox Vision and the Christian Right*. Chicago: University of Chicago Press.

Herrnstein, Richard, and Charles Murray. 1994. *The Bell Curve: Intelligence and Class Structure in American Life*. New York: The Free Press.

Hicks, Michael. 1991. "The Imprisonment of Henry Cowell." *Journal of the American Musicological Society* 44:92–119.

——. 1993. "Cowell's Clusters." *The Musical Quarterly* 77:428–58.

——. 2005. *Henry Cowell: Bohemian*. Urbana: University of Illinois Press.

Hird, Myra. 2004. "Naturally Queer." *Feminist Theory* 5:85–89.

Hobbs, A. H., and R. D. Lambert. 1948. "An Evaluation of 'Sexual Behavior in the Human Male.'" *The American Journal of Psychiatry* 104:758–64.

Hollingworth, H. 1939. "Review of 'Psychological Factors in Marital Happiness.'" *Psychological Bulletin* 36:191–97.

Hollingworth, L. S. 1916. "Social Devices for Impelling Women to Bear and Rear Children." *American Journal of Sociology* 1:19–29.

Hollingworth, Leta Stetter, and M. M. Rust. 1937. "Application of the Bernreuter Inventory of Personality to Highly Intelligent Adolescents." *Journal of Psychology: Interdisciplinary and Applied* 4:287–93.

Hoogland, David Noon. 2004. "Situating Gender and Professional Identity in American Child Study, 1880–1910." *History of Psychology* 7:107–29.

Horney, Karen. 1926. "The Flight from Womanhood: The Masculinity Complex in Women as Viewed by Men and by Women." *International Journal of Psychoanalysis* 7:324–39.

Hornstein, Gail A. 1988. "Quantifying Psychological Phenomena: Debates, Dilemmas and Implications." In *The Rise of Experimentation in American Psychology*, ed. Jill G. Morawski, 1–34. New Haven: Yale University Press.

Hook, Derek. 2001. "Discourse, Knowledge, Materiality, History: Foucault and Discourse Analysis." *Theory & Psychology* 11:521–47.

——. 2003. "Analogies of Power: Reading Psychotherapy through the Sovereignty-Discipline-Government Complex." *Theory & Psychology* 13:605–28.

Hotelling, Harold. 1927. "Review of Ronald Fisher's Statistical Methods for Research Workers." *Journal of the American Statistical Association* 22:411–12.

Hubbs, Nadine. 2004. *The Queer Composition of America's Sound: Modernists, American Music, and National Identity*. Durham, N.C.: Duke University Press.

Huertas, R. 1993. "Madness and Degeneration, IV: The Man of Genius." *History of Psychiatry* 4:301–19.

Hyman, Herbert H., and Paul B. Sheatsley. 1948. "The Kinsey Report and Survey Methodology." *International Journal of Opinion and Attitudes Research* 2:183–95.

Igo, Sarah E. 2006. "'A Gold Mine and a Tool for Democracy': George Gallop, Elmo Ropper and the Business of Scientific Polling, 1935–1955." *Journal of the History of the Behavioral Sciences* 42:109–34.

———. 2007. *The Averaged American: Surveys, Citizens and the Making of a Mass Public.* Cambridge, Mass.: Harvard University Press.

Irvine, Janice. 1990. *Disorders of Desire: Sex and Gender in Modern American Sexology.* Philadelphia: Temple University Press.

———. 2002. *Talk About Sex: The Battles Over Sex Education in the United States.* Berkeley and Los Angeles: University of California Press.

Johnston, Elizabeth, and Ann Johnson. 2008. "Searching for the Second Generation of American Women Psychologists." *History of Psychology* 11:40–72.

Jones, James H. 1997. *Alfred C. Kinsey: A Public/Private Life.* New York: W. W. Norton & Co.

Jorgensen, Elizabeth Watkins, and Henry I. Jorgensen. 1999. *Thorstein Veblen: Victorian Firebrand.* Armonk, N.Y.: M. E. Sharpe, Inc.

Katz, Jonathan Ned. 1976. *Gay American History: Lesbians and Gay Men in the USA: A Documentary History.* New York: Meridian.

Keller, Evelyn Fox. 1985. *Reflections on Gender and Science.* New Haven: Yale University Press.

———. 1992. *Secrets of Life, Secrets of Death: Essays on Language, Gender and Science.* New York: Routledge.

Kelly, E. Lowell. 1939. "Concerning the Validity of Terman's Weights for Predicting Marital Happiness." *Psychological Bulletin* 36:202–3.

———. 1950. "Consistency of the Adult Personality." *American Psychologist* 10:659–81.

Kelly, E. Lowell, and James Conley. 1987. "Personality and Compatibility: A Prospective Analysis of Marital Stability and Marital Satisfaction." *Journal of Personality and Social Psychology* 52:27–40.

Kelman, Herbert C. 1967. "Human Use of Human Subjects: The Problem of Deception in Social Psychological Experiments." *Psychological Bulletin* 67:1–11.

Kendall, Lori. 1999. "Nerd Nation: Images of Nerds in US Popular Culture." *International Journal of Cultural Studies* 2:260–83.

Kevles, Daniel J. 1968. "Testing the Army's Intelligence." *Journal of American History* 55:565–81.

———. 1985. *In the Name of Eugenics: Genetics and the Use of Human Heredity.* New York: Alfred A. Knopf, Inc.

Kevles, Daniel J., and Leroy Hood. 1993. *The Code of Codes: Scientific and*

Social Issues in the Human Genome Project. Cambridge, Mass.: Harvard University Press.

King, D. Brett, Lizzi M. Montanez-Ramirez, and Michael Wertheimer. 1996. "Barbara Stockard Burks: Pioneer Behavioral Geneticist and Humanitarian." In *Portraits of Pioneers in Psychology*, vol. 12, ed. Gregory A. Kimble, C. Alan Boneau, and Michael Wertheimer, 213–25. Washington, D.C.: American Psychological Association.

King, Willford I. 1936. "Consolidating Our Gains." *Journal of the American Statistical Association* 31:1–9.

Kingsland, Sharon. 1995. *Modeling Nature: Episodes in the History of Population Ecology.* 2d ed. Chicago: University of Chicago Press.

Kinsey, Alfred C. 1923. "The Gall Wasp Genus Neuroterus (Hymenoptera)." *Indiana University Studies* 10:3–150.

———. 1929. "The Gall Wasp Genus Cynips: A Study in the Origin of Species." *Indiana University Studies* 16:1–577.

———. 1935. "The Economic Importance of the Cynipidae." *Journal of Economic Entomology* 28:86–91.

———. 1937. "An Evolutionary Analysis of Insular and Continental Species." *Proceedings of the National Academy of Sciences of the United States of America* 23:5–11.

———. 1941. "Criteria for a Hormonal Explanation of the Homosexual." *Journal of Clinical Endocrinology* 1:424–28.

———. "Individuals." 1971. In *Kinsey: A Biography*, by Cornelia V. Christiansen, 3–9. Bloomington: Indiana University Press.

Kinsey, Alfred C., Wardell B. Pomeroy, and Clyde E. Martin. 1948. *Sexual Behavior in the Human Male.* Philadelphia: W. B. Saunders and Company.

Kinsey, Alfred C., Wardell B. Pomeroy, Clyde E. Martin, and Paul H. Gebhard. 1953. *Sexual Behavior in the Human Female.* Philadelphia: W. B. Saunders and Company.

Kinsey, Alfred C., et al. 1955. "The Cochran-Mosteller-Tukey Report on the Kinsey Study: A Symposium." *Journal of the American Statistical Association*, 50.271: 811–29.

Kissack, Terrence. 2000. "Alfred Kinsey and Homosexuality in the '50s: The Recollections of Samuel Morris Steward As Told To Len Evans." *Journal of the History of Sexuality* 9:474–91.

Kitzinger, Celia. 1987. *The Social Construction of Lesbianism.* London: Sage.

Kline, Wendy. 2001. *Building a Better Race: Gender, Sexuality and Eugenics from the Turn of the Century to the Baby Boom.* Berkeley and Los Angeles: University of California Press.

Kohler, Robert E. 2002. *Landscapes and Labscapes: Exploring the Lab-Field Border in Biology*. Chicago: University of Chicago Press.

Krantz, D. L. 2001. "Reconsidering History of Psychology's Borders." *History of Psychology* 4:182–94.

Kruger, Lorenz, Lorraine J. Daston, and Michael Heiderberger. 1987. *The Probabilistic Revolution*, vol. 1: *Ideas in History*. Cambridge, Mass.: MIT Press.

Kruger, Lorenz, Gerd Gigerenzer, and Mary S. Morgan. 1987. *The Probabilistic Revolution*, vol. 2: *Ideas in the Sciences*. Cambridge, Mass.: MIT Press.

Kühl, Stefan. 2001. *The Nazi Connection: Eugenics, American Racism and German National Socialism*. New York: Oxford University Press.

Kuhn, Thomas. 1970. *The Structure of Scientific Revolutions*. 2d ed. Chicago: University of Chicago Press.

Ladd-Taylor, Molly. 2001. "Eugenics, Sterilization and Modern Marriage in the USA: The Strange Career of Paul Popenoe." *Gender & History* 13:298–327.

Landis, Carney. 1948. "Book Review: Kinsey, Pomeroy, and Martin, *Sexual Behavior in the Human Male*." *Psychological Bulletin* 45:272–75.

Landis, Carney, and Mary Marjorie Bolles. 1942. *Personality and Sexuality of the Physically Handicapped Woman*. New York: Paul B. Hoeber.

Laqueur, Thomas W. 2003. *Solitary Sex: A Cultural History of Masturbation*. New York: Zone Books.

Latour, Bruno. 1983. "Give Me a Laboratory and I Will Raise the World." In *Science Observed: Perspectives on the Social Study of Science*, ed. Karin D. Knorr-Cetina and Michael Mulkay, 141–70. London: Sage.

———. 1987. *Science in Action: How to Follow Scientists and Engineers through Society*. Cambridge, Mass.: Harvard University Press.

———. 1990. "Drawing Things Together." In *Representation in Scientific Practice*, ed. Michael Lynch and Steve Woolgar, 19–68. Cambridge, Mass.: MIT Press.

———. 1993. *We Have Never Been Modern*. Translated by Catherine Porter. Cambridge, Mass.: Harvard University Press.

Latour, Bruno, and Steve Woolgar. 1985. *Laboratory Life: The Construction of Scientific Facts*. 2d ed. Princeton: Princeton University Press.

Laumann, Edward O., John H. Gagnon, Robert T. Michael, and Stuart Michaels. 1994. *The Social Organization of Sexuality*. Chicago: University of Chicago Press.

Lemov, Rebecca. 2005. *World as Laboratory: Experiments with Mice, Mazes and Men*. New York: Hill & Wang.

———. 2011. "X-rays of Inner Worlds: The Mid-Twentieth-Century American Projective Test Movement." *Journal of the History of the Behavioral Sciences* 47:251–78.

Levine, J. 2003. *Harmful to Minors: The Perils of Protecting Children from Sex.* New York: Da Capo Press.

Lewin, Miriam. 1984. "Psychology Measures Femininity and Masculinity, 2: From '13 Gay Men' to the Instrumental-Expressive Distinction." In *In the Shadow of the Past: Psychology Portrays the Sexes: A Social and Intellectual History*, ed. Miriam Lewin, 179–204. New York: Columbia University Press.

Lippa, Richard A. 2005. *Gender, Nature and Nurture.* 2d ed. Mahwah, N.J.: Psychology Press.

Lippa, Richard A., Leslie R. Martin, and Howard S. Friedman. 2000. "Gender-Related Individual Differences and Mortality in the Terman Longitudinal Study: Is Masculinity Hazardous for Your Health?" *Personality and Social Psychology Bulletin* 26:1560–70.

Logan, William Bryant. 2006. *Oak: The Frame of Civilization.* New York: W. W. Norton.

Lombardo, Giovanni Pietro, and Renato Forschi. 2003. "The Concept of Personality in 19th-Century French and 20th-Century American Psychology." *History of Psychology* 6:123–42.

Lombroso, Cesare. 1889. *The Man of Genius.* London: Scott.

Lorde, Audre. 1984. *Sister Outsider: Essays and Speeches by Audre Lorde.* Berkeley and Toronto: Crossing Press.

Lowen, Rebecca. 1997. *Creating the Cold War University: The Transformation of Stanford.* Berkeley and Los Angeles: University of California Press.

Lubek, Ian. 2000. "Understanding and Using the History of Social Psychology." *Journal of the History of the Behavioral Sciences* 36:499–506.

Lubek, Ian, and Hendrickson J. Stam. 1995. "Ludicro-Experimentation in Social Psychology: Sober Scientific versus Playful Prescriptions." In *Trends and Issues in Theoretical Psychology*, ed. I. Lubek, R. van Hezewijk, G. Pheterson and C. W. Tolman, 171–80. New York: Springer.

Lubin, Isador. 1947. "Social and Economic Adjustments in a Democratic World." *Journal of the American Statistical Association* 42:11–19.

Lunbeck, Elizabeth. 1995. *The Psychiatric Persuasion: Knowledge, Gender and Power in Modern America.* Princeton: Princeton University Press.

Maines, Rachel P. 2001. *The Technology of Orgasm: "Hysteria," the Vibrator and Women's Sexual Satisfaction.* Baltimore: Johns Hopkins University Press.

Malthus, Thomas Robert. 1798. *An Essay on the Principle of Population.* London: J. Johnson.

Manning, J. T., S. Baron-Cohen, S. Wheelwright, and G. Sanders. 2001. "The 2nd to 4th Digit Ratio and Autism." *Developmental Medicine & Child Neurology* 43:160–64.

Margolin, Leslie. 1993. "Goodness Personified: The Emergence of Gifted Children." *Social Problems* 40:510–32.

———. 1994. *Goodness Personified: The Emergence of Gifted Children.* New York: Aldine de Gruyter.

Martin, William C. 2005. *With God On Our Side: The Rise of the Religious Right in America.* New York: Broadway Books.

Maslow, Abraham H., and James M. Sakoda. 1952. "Volunteer-Error in the Kinsey Study." *Journal of Abnormal and Social Psychology* 47:259–62.

Mauss, Marcel. [1924] 2002. *The Gift.* London: Routledge Classics.

May, Elaine Tyler. 1988. *Homeward Bound: American Families in the Cold War Era.* New York: Basic Books.

May, Geoffrey. 1931. *Social Control of Sex Expression.* New York: William Morrow & Co.

Mayr, Ernst. 1942. *Systematics and the Origins of Species: From the Viewpoint of a Zoologist.* New York: Columbia University Press.

———. 1982. *The Growth of Biological Thought.* Cambridge, Mass.: Harvard University Press.

McCaughey, M. 2008. *The Caveman Mystique: Pop Darwinism and the Debates Over Sex, Violence, and Science.* New York: Routledge.

McConahay, John B. 1983. "Modern Racism and Modern Discrimination: The Effects of Race, Racial Attitudes, and Context on Simulated Hiring Decisions." *Personality and Social Psychology Bulletin* 9:551–58.

McCullough, Michael, Jo-Ann Tsang, and Sharon Brion. 2003. "Personality Traits in Adolescence as Predictors of Religiousness in Early Adulthood: Findings from the Terman Longitudinal Study." *Personality and Social Psychology Bulletin* 29:980–91.

McNemar, Quinn. 1940. "Sampling in Psychological Research." *Psychological Bulletin* 37:331–65.

———. 1946. "Opinion-Attitude Methodology." *Psychological Bulletin* 43:289–374.

———. 1980. "Quinn McNemar." In *A History of Psychology in Autobiography*, vol. 7, ed. Garnder Lindsey, 305–33. San Francisco: Freeman & Co.

McNemar, Quinn, and Lewis M. Terman. 1936. "Sex Differences in Variational Tendency." *Genetic Psychology Monographs* 18:1–65.

Mills, Frederick C. 1935. "Statistics and Leviathan." *Journal of the American Statistical Association* 30:1–11.

Mills, John A. 1998. *Control: A History of Behavioral Psychology.* New York: New York University Press.

Minton, Henry L. 1984. "The Iowa Child Welfare Research Station and the 1940 Debate on Intelligence: Carrying on the Legacy of a Concerned Mother." *Journal of the History of the Behavioral Sciences* 20:160–74.

———. 1988. *Lewis M. Terman: Pioneer in Psychological Testing.* New York: New York University Press.

———. 2002. *Departing from Deviance: A History of Homosexual Rights and Emancipatory Science in America.* Chicago: University of Chicago Press.

Moghaddam, Fathali M., and Naomi Lee. 2006. "Double Reification: The Process of Universalizing Psychology in the Three Worlds." In *Internationalizing the History of Psychology*, ed. Adrian C. Brock, 163–82. New York: New York University Press.

Moore, Lisa Jean. 2002. "Extracting Men from Semen: Masculinity in Scientific Representations of Sperm." *Social Text* 20:91–119.

Moran, Jeffrey P. 2000 *Teaching Sex: The Shaping of Adolescence in the 20th Century.* Cambridge, Mass.: Harvard University Press.

Morantz, Regina Markell. 1977. "The Scientist as Sex Crusader: Alfred C. Kinsey and American Culture." *American Quarterly* 29:563–89.

Morawksi, Jill G. 1992. "Self-Regard and Other-Regard: Reflexive Practices in Psychology, 1890–1940." *Science in Context* 5:281–308.

———. 1994. *Practicing Feminisms, Reconstructing Psychology: Notes on a Liminal Science.* Ann Arbor: University of Michigan Press.

———. 2005. "Reflexivity and the Psychologist." *History of the Human Sciences* 18:77–105.

Morawksi, Jill G., ed. 1988. *The Rise of Experimentation in American Psychology.* New Haven: Yale University Press.

Morokoff, Patricia J. 1986. "Volunteer Bias in the Psychophysiological Study of Female Sexuality." *Journal of Sex Research* 22.1:35–51.

Morrison, Melanie, and Todd Morrison. 2002. "Development and Validation of a Scale Measuring Modern Prejudice Toward Gay Men and Lesbian Women." *Journal of Homosexuality* 43:15–37.

Mulder, Ernst, and Frieda Heyting. 1998. "The Dutch Curve: The Introduction and Reception of Intelligence Testing in the Netherlands, 1908–1940." *Journal of the History of the Behavioral Sciences* 34:349–66.

Muschinske, David. 1977. "The Nonwhite as Child: G. Stanley Hall on the

Education of Nonwhite Peoples." *Journal of the History of the Behavioral Sciences* 13:328–36.

Napoli, Donald S. 1980. *Architects of Adjustment: The History of the Psychological Profession in the United States*. Port Washington, N.Y.: Associated Faculty Press Inc.

Neisser, Ulrich, Gwyneth Boodoo, Thomas J. Bouchard Jr, A. Wade Boykin, Nathan Brody, Stephen J. Ceci, Diane F. Halperin, John C. Loehlin, Robert Perloff, Robert J. Sternberg, and Susana Urbina. 1996. "Intelligence: Knowns and Unknowns." *American Psychologist* 51:77–101.

Nelkin, Dorothy, and M. Susan Lindee. 1995. *The DNA Mystique: The Gene as a Cultural Icon*. San Francisco: W. H. Freeman & Co.

Neuhaus, Jessamyn. 2000. "The Importance of Being Orgasmic: Sexuality, Gender and Marital Sex Manuals in the United States, 1920–1963." *Journal of the History of Sexuality* 9:447–73.

Neumann, R. P. 1975. "Masturbation, Madness, and the Modern Concepts of Childhood and Adolescence." *Journal of Social History* 8:1–27.

Nicholson, Ian A. M. 2003. *Inventing Personality: Gordon Allport and the Science of Selfhood*. Washington D.C.: American Psychological Association.

Nordau, Max. [1892] 2006. *Degeneration*. Lincoln: University of Nebraska Press.

O'Connell, Agnes N., and Nancy Filipe Russo. 1990. *Women in Psychology: A Bio-Bibliographic Sourcebook*. New York: Greenwood Press.

O'Donnell, John M. 1985. *The Origins of Behaviorism: American Psychology 1870–1920*. New York: New York University Press.

Olbricht, Erika Mae. 2006. "Made without Hands: The Representation of Labor in Early Modern Silkworm and Beekeeping Manuals." In *Insect Poetics*, ed. E. C. Brown, 223–41. Minneapolis: University of Minnesota Press.

Ordover, Nancy. 2003. *American Eugenics: Race, Queer Anatomy and the Science of Nationalism*. Minneapolis: University of Minnesota Press.

Ott, Katherine. 1996. *Fevered Lives: Tuberculosis in American Culture since 1870*. Cambridge, Mass.: Harvard University Press, 1996.

Oyama, Susan. 2000. *The Ontogeny of Information: Developmental Systems and Evolution*. 2d ed. Durham, N.C.: Duke University Press.

Passet, Joanne. 2008. *Sex Variant Woman: The Life of Jeanette Howard Foster*. New York: Avalon.

Pauly, Philip J. 2002. *Biologists and the Promise of American Life: Meriwether Lewis to Alfred Kinsey*. Princeton: Princeton University Press.

Pearl, Raymond. 1922. *The Biology of Death*. Philadelphia: J. B. Lippincott and Co.

Pickren, Wade. 1997. "Robert Yerkes, Calvin Stone and the Beginnings of Programmatic Sex Research by Psychologists, 1921–1930." *American Journal of Psychology* 110:603–19.

Plotz, David. 2005. *The Genius Factory: Unravelling the Mysteries of the Nobel Prize Sperm Bank*. London: Simon and Schuster.

Pomeroy, Wardell B. 1972. *Dr. Kinsey and the Institute for Sex Research*. London: Thomas Nelson & Sons.

Popenoe, Paul. 1925. *Modern Marriage: A Handbook*. New York: The Macmillan Company.

———. 1934. "The German Sterilization Law." *Journal of Heredity* 25:257–60.

Popper, Deborah Epstein, Robert E. Lang, and Frank J. Popper. 2000. "From Maps to Myth: The Census, Turner, and the Idea of the Frontier." *The Journal of American Culture* 23:91–102.

Popper, Karl. 1959. *The Logic of Scientific Discovery*. London: Hutchinson & Co.

Potter, Jonathan, and Margaret Wetherell. 1987. *Discourse and Social Psychology: Beyond Attitudes and Behaviour*. London: Sage.

Privateer, Paul M. 2005. *Inventing Intelligence: A Social History of Smart*. Oxford Wiley-Blackwell.

Pruitt, Matthew V. 2002. "Size Matters: A Comparison of Anti- and Pro-Gay Organizations' Estimates of the Size of the Gay Population." *Journal of Homosexuality* 42:21–29.

Pullias, E. V. 1937. "Masturbation as a Mental Hygiene Problem: A Study of the Beliefs of Seventy-Five Young Men." *The Journal of Abnormal and Social Psychology* 32:216–22.

Ramsden, Edmund. 2002. "Carving Up Population Sciences: Eugenics, Demography, and the Controversy over the 'Biological Law' of Population Growth." *Social Studies of Science* 32:857–99.

———. 2007. "A Differential Paradox: The Controversy Surrounding the Scottish Mental Surveys of Intelligence and Family Size." *Journal of the History of the Behavioral Sciences* 43:109–34.

Ramsey, Glenn V. 1943. "The Sex Information of Younger Boys." *American Journal of Orthopsychiatry* 13:347–52.

Reed, James. 1983. *The Birth Control Movement and American Society: From Private Vice to Public Virtue*. Princeton: Princeton University Press.

Reumann, Miriam. 2005. *American Sexual Character: Sex, Gender, and Na-*

tional Identity in the Kinsey Reports. Berkeley and Los Angeles: University of California Press.

Rice, Stuart A. 1934. "Statistical Opportunities and Responsibilities." *Journal of the American Statistical Association* 29:1–10.

Richards, Graham. 1995. "'To Know Our Fellow Men To Do Them Good': American Psychology's Enduring Moral Project." *History of the Human Sciences* 8:1–24.

———. 1997. *Race, Racism and Psychology: Toward a Reflexive History*. London: Routledge.

———. 2004. "'It's an American Thing': The 'Race' and Intelligence Controversy from a British Perspective." In *Defining Difference: Race and Racism in the History of Psychology*, ed. Andrew S. Winston, 137–69. Washington, D.C.: American Psychological Association.

———. 2010. *Putting Psychology in its Place: An Introduction from a Critical Historical Perspective*. 3d ed. London: Routledge.

———. 2011. *Psychology, Religion, and the Nature of the Soul: A Historical Entanglement*. Tunbridge Wells: Graham Richards Books.

Richardson, Ken. 2002. "What IQ Tests Test." *Theory & Psychology* 12:283–314.

Rieger, Gerulf, Joan A. W. Linsenmeier, and J. Michael Bailey. 2009. "Childhood Gender Nonconformity Remains a Robust and Neutral Correlate of Sexual Orientation: Reply to Hegarty (2009)." *Developmental Psychology* 45:901–3.

Roberts, Dorothy E. 1997. *Killing the Black Body: Race, Reproduction and the Meaning of Liberty*. New York: Pantheon.

Robinson, Paul A. 1977. *The Modernization of Sex: Havelock Ellis, Alfred Kinsey, William Masters, and Virginia Johnson*. New York: Harper & Row.

Rogers, Naomi. 1989. "Dirt, Flies and Immigrants: Explaining the Epidemiology of Poliomyelitis, 1900–1916." *Journal of the History of Medicine and Allied Sciences* 44:486–505.

Rose, Nikolas. 1996. *Inventing Our Selves: Psychology Power and Personhood*. Cambridge: Cambridge University Press.

Rosenberg, Carroll Smith. 1985. *Disorderly Conduct: Visions of Gender in Victorian America*. New York: Oxford University Press.

Rosenman, Ellen Bayuk. 2003. "Body Doubles: The Spermatorrhea Panic." *Journal of the History of Sexuality* 12:365–99.

Rosenzweig, Louise, and Saul Rosenzweig, S. 1969. "Notes on Alfred C. Kinsey's Presexual Scientific Work and the Transition." *Journal of the History of the Behavioral Sciences* 5:173–81.

Rosenzweig, Saul. 1992. *Freud, Jung and Hall the King-Maker: The Historical Expedition to America (1909)*. St. Louis: Rana House.

Ross, Andrew. 1989. *No Respect: Intellectuals and Popular Culture*. New York: Routledge.

Ross, Dorothy. 1972. *G. Stanley Hall: The Psychologist as Prophet*. Chicago: University of Chicago Press.

Ross, Robert T. 1950. "Measures of the Sex Behaviour of College Males Compared with Kinsey's Results." *The Journal of Abnormal and Social Psychology* 45:753–55.

Roth, Wolff-Michael. 2003. *Toward an Anthropology of Graphing: Semiotic and Activity-Theoretic Perspectives*. Dordrecht: Kluwer Academic Publishers.

Rothman, Sheila W. 1994. *Living in the Shadow of Death: Tuberculosis and the Social Experience of Illness in American History*. New York: Basic Books.

Rousseau, Jean-Jacques. 1963. *Emile*. Translated by Grace G. Roosevelt. London: Everyman's Library.

Royse, N. K. 1891. *A Study of Genius*. New York: Rand, McNally and Co.

Rucci, Anthony J., and Ryan D. Tweney. 1980. "Analysis of Variance and the 'Second Discipline' of Scientific Psychology: A Historical Analysis." *Psychological Bulletin* 87:166–84.

Rushton, J. Philippe. 2000. *Race, Evolution and Behavior: A Life History Perspective*. 2d special abridged ed. Port Huron, Mich.: Charles Darwin Research Institute.

Rushton, J. Philippe, and Arthur Jensen. 2005. "Thirty Years of Research on Race and IQ." *Psychology, Public Policy and Law* 11:235–94.

Russell, Edmund P. 1996. "'Speaking of Annihilation': Mobilizing for War Against Human and Insect Enemies, 1914–1945." *Journal of American History* 82:1505–29.

Rutherford, Alexandra. 2009. *Beyond the Box: B. F. Skinner's Technology of Behavior from Laboratory to Life, 1950s to 1970s*. Toronto: University of Toronto Press.

Said, Edward. 2004. *Freud and the Non-European*. London: Verso.

Samelson, Franz. 1978. "From 'Race Psychology' to 'Studies in Prejudice': Some Observations on the Thematic Reversal in Social Psychology." *Journal of the History of Behavioral Sciences* 14:265–78.

———. 2000. "Whig and Anti-Whig Histories and Other Curiosities of Social Psychology." *Journal of the History of the Behavioral Sciences* 36:499–506.

Sandfort, Theo G. M. 2005. "Sexual Orientation and Gender: Stereotypes and Beyond." *Archives of Sexual Behavior* 34:595–611.

Sanger, Margaret. 1919. "Birth Control and Racial Betterment." *Birth Control Review* 3:1–12.

———. 1938. *Margaret Sanger: An Autobiography*. New York: W. W. Norton & Co.

———. [1921] 2002. "The Eugenic Value of Birth Control Propaganda." *The Selected Papers of Margaret Sanger*, vol. 1: *The Woman Rebel, 1900–1928*. Urbana: University of Illinois Press.

Sasson, Ralph, and T. M. Nelson. 1969. "The Human Experimental Subject in Context." *Canadian Psychologist* 10:409–37.

Sawyer, Thomas F. 2000. "Francis Cecil Sumner: His Views and Influence on African American Higher Education." *History of Psychology* 3:122–41.

Schmader, Toni, Michael Johns, and Chad Forbes. 2008. "An Integrated Process Model of Stereotype Threat Effects on Performance." *Psychological Review* 115:336–56.

Schneider, William H. 1992. "After Binet: French Intelligence Testing, 1900–1950." *Journal of the History of the Behavioral Sciences* 28:111–32.

Scully, Diana, and Pauline Bart. 1973. "A Funny Thing Happened on the Way to the Orifice: Women in Gynaecology Textbooks." *American Journal of Sociology* 78:1045–50.

Seagoe, May V. 1975. *Terman and the Gifted*. Los Altos, Calif.: William Kaufman, Inc.

Sears, David O. 1986. "College Sophomores in the Laboratory: Influences of a Narrow Data Base on Social Psychology's View of Human Nature." *Journal of Personality and Social Psychology* 51:515–30.

Sears, Pauline S., and A. H. Barbee. 1977. "Career and Life Satisfaction among Terman's Gifted Women." In *The Gifted and the Creative: A Fifty Year Perspective*, ed. Julian C. Stanley, William C. George, and Cecilia H. Solano, 28–65. Baltimore: Johns Hopkins University Press.

Sedgwick, Eve Kosovsky. 1990. *Epistemology of the Closet*. Berkeley and Los Angeles: University of California Press.

Serlin, David. 2012. "Carney Landis and the Psychosexual Landscape of Touch in Mid-20th Century America." *History of Psychology* 15:209–16.

Shapin, Steven. 1995. *A Social History of Truth: Civility and Science in Seventeenth-Century England*. Chicago: University of Chicago Press.

———. 1996. *The Scientific Revolution*. Chicago: University of Chicago Press.

———. 2008. *The Scientific Life: A Moral History of a Late Modern Vocation*. Chicago: University of Chicago Press.

Shewart, Walter A. 1946. "The Advancing Statistical Front." *Journal of the American Statistical Association* 41:1–15.

Shields, Stephanie A. 1982. "The Variability Hypothesis: The History of a Biological Model of Sex Differences in Intelligence." *Signs* 7:769–97.

———. 1991. "Leta Stetter Hollingworth: 'Literature of Opinion' and the Study of Individual Differences." In *Portraits of Pioneers in Psychology*, ed. George A. Kimble, Michael Wertheimer, and Charlotte White, 243–55. Hillsdale, N.J.: Erlbaum.

Shurkin, Joel L. 1992. *Terman's Kids: The Groundbreaking Study of How the Gifted Grow Up*. New York: Little, Brown and Co.

Siegler, Robert S. 1992. "The Other Alfred Binet." *Developmental Psychology* 28:179–90.

Siegman, Aron W. 1956. "Responses to a Personality Questionnaire by Volunteers and Nonvolunteers to a Kinsey Interview." *Journal of Abnormal and Social Psychology* 52:280–81.

Simonton, Dean Keith. 1999. *Origins of Genius: Darwinian Perspectives on Creativity*. New York: Oxford University Press.

———. 2009. "The 'Other IQ': Historiometric Assessments of Intelligence and Related Constructs." *Review of General Psychology* 13:315–26.

Sleigh, Charlotte. 2002. "Brave New Worlds: Trophylaxis and the Origin of Society in the Early 20th Century." *Journal of the History of the Behavioral Sciences* 38:133–56.

———. 2006. "Inside Out: The Unsettling Nature of Insects." In *Insect Poetics*, ed. E. C. Brown, 281–97. Minneapolis: University of Minnesota Press.

———. 2007. *Six Legs Better: A Cultural History of Myrmecology*. Baltimore: Johns Hopkins University Press.

Smith, Laurence D. 1986. *Behaviorism and Logical Positivism: A Reassessment of the Alliance*. Stanford: Stanford University Press.

Smith, Roger. 1988. "Does the History of Psychology Have a Subject?" *History of the Human Sciences* 1:147–77.

Smith-Rosenberg, Carroll. 1985. *Disorderly Conduct: Visions of Gender in Victorian America*. New York: Oxford University Press.

Smuts, Alice Boardman. 2006. *Science in the Service of Children: 1893–1935*. New Haven: Yale University Press.

Sokal, Michael. 1990. "G. Stanley Hall and the Institutional Character of Psychology at Clark, 1889–1920." *Journal of the History of the Behavioral Sciences* 26:114–24.

Spearman, Charles. 1904. "General Intelligence Objectively Determined and Measured." *American Journal of Psychology* 15:201–93.

———. 1930. "Charles Spearman." In *A History of Psychology in Autobiography*, vol. 1, ed. Carl Murchison. Worchester, Mass.: Clark University Press.

Steele, Claude, and Joshua Aronson. 1995. "Stereotype Threat and the Intellectual Test Performance of African Americans." *Journal of Personality and Social Psychology* 69:797–811.

Stengers, Jean, and Anne Van Neck. [1984] 2001. *Masturbation: The History of a Great Terror*. Translated by Kathryn Hoffman. New York: Palgrave.

Stephan, Frederick F. 1939. "Representative Sampling in Large-Scale Surveys." *Journal of the American Statistical Association* 34:343–52.

Stern, Alexandra Minna. 2005. *Eugenic Nation: Faults and Frontiers of Better Breeding in Modern America*. Berkeley and Los Angeles: University of California Press.

Stewart, Alan. 2003. "Homosexuals in History: A. L. Rowse ad the Queer Archive." In *Love, Sex, Intimacy and Friendship Between Men, 1550–1800*, ed. Katherine O'Donnell and Michael O'Rourke, 53–69. Basingstoke, U.K.: Palgrave Macmillan.

Stigler, Stephen M. 1989. "A Look Backward on the Occasion of the Centenary of JASA." *Journal of the American Statistical Association* 83:583–87.

Stipanovic, Andre. 2006. "Bees and Ants: Perceptions of Imperialism in Vergil's *Aeneid* and *Georgics*." In *Insect Poetics*, ed. E. C. Brown, 13–28. Minneapolis: University of Minnesota Press.

Stocking, George W. 1965. "On the Limits of 'Presentism' and 'Historicism' in the Historiography of the Behavioral Sciences." *Journal of the History of the Behavioral Sciences* 1:211–18.

Storfer, Miles D. 1990. *Intelligence and Giftedness: The Constitutions of Heredity and Early Environment*. San Francisco: Jossey-Bass.

Strong, Edward K. 1948. *Vocational Interests of Men and Women*. Stanford: Stanford University Press.

Swim, Janet K., Kathryn J. Aikin, Wayne S. Hall, and Barbara A. Hunter. 1995. "Sexism and Racism: Old-Fashioned and Modern Prejudices." *Journal of Personality and Social Psychology* 68:199–214.

Teller, Michael E. 1988. *The Tuberculosis Movement: A Public Health Campaign in the Progressive Era*. New York: Greenwood Press.

Terry, Jennifer. 1999. *An American Obsession: Science, Medicine and Homosexuality in Modern Society*. Chicago: University of Chicago Press.

Terman, Lewis M. 1905. "A Study in Precocity and Prematuration." *American Journal of Psychology* 16:145–83.

———. 1906. "Genius and Stupidity: A Study of Some of the Intellectual Processes of Seven 'Bright' And Seven 'Stupid' Boys." *Pedagogical Seminary* 13:307–73.

———. 1915. "The Mental Hygiene of Exceptional Children." *Pedagogical Seminary* 22:529–37.

———. 1916. *The Measurement of Intelligence*. Boston: Houghton Mifflin.

———. 1917. "The Intelligence Quotient of Francis Galton in Childhood." *American Journal of Psychology* 28:209–15.

———. 1921. *The Intelligence of School Children*. London: Harrap.

———. 1925. *Genetic Studies of Genius*, vol. 1: *Mental and Physical Traits of One Thousand Gifted Children*. Stanford: Stanford University Press.

———. 1930. "Trails to Psychology: Autobiography of Lewis M. Terman." In *History of Psychology in Autobiography*, vol. 2, ed. Carl Murchison, 297–331. Worcester, Mass: Clark University Press.

———. 1938. *Psychological Factors in Marital Happiness*. New York: McGraw-Hill.

———. 1939. "The Effect of Happiness or Unhappiness on Self-Report Regarding Attitude Reaction Patterns and Facts of Personal History." *Psychological Bulletin* 36:197–202.

———. 1941. "Should the Historian Study Psychology?" *Pacific Historical Review* 10:209–16.

———. 1948. "Kinsey's 'Sexual Behavior in the Human Male': Some Comments and Criticisms." *Psychological Bulletin* 45:443–59.

Terman, Lewis M., N. Bayley, Helen Marshall, Olga W. McNemar, and Melita H. Oden. 1951. "Correlates of Orgasm Adequacy in a Group of 556 Wives." *Journal of Psychology: Interdisciplinary and Applied* 32:115–72.

Terman, Lewis M., Barbara S. Burks, and D. W. Jensen. 1930. *Genetic Studies of Genius*, vol. 3: *The Promise of Youth: Follow-up Studies of One Thousand Gifted Children*. Stanford: Stanford University Press.

Terman, Lewis M., and Catherine Cox Miles. 1936. *Sex and Personality: Studies in Masculinity and Femininity*. New York: McGraw-Hill.

Terman, Lewis M., and Melita H. Oden. 1947. *Genetic Studies of Genius*, vol. 4: *The Gifted Child Grows Up: Twenty-Five Years' Follow-up of a Superior Group*. Stanford: Stanford University Press.

———. 1959. *Genetic Studies of Genius*, vol. 5: *The Gifted Child at Mid-Life: Thirty-Five Years' Follow-up of the Superior Child*. Stanford: Stanford University Press.

Thagard, Paul. 2006. *Hot Thought: Mechanisms and Applications of Emotional Cognition*. Cambridge, Mass.: MIT Press.

Thorp, Willard. 1948. "Statistics and Foreign Policy." *Journal of the American Statistical Association* 43:1–11.

Tilman, Rick. 1996. *The Intellectual Legacy of Thorstein Veblen: Unresolved Issues*. Westport, Conn.: Greenwood Press.

Trilling, Lionel. 1954. "The Kinsey Report." In *An Analysis of the Kinsey Reports on Sexual Behavior in the Human Male and Female*, ed. Donald Porter Geddes, 213–29. New York: Mentor.

Turing, Alan M. 1950. "Computing Machinery and Intelligence." *Mind* 54:433–60.

Turner, Frederick Jackson. [1920] 2011. *The Frontier in American History*. CreateSpace. Originally published New York: H. Holt and Company.

United Nations. 1948. *The Universal Declaration of Human Rights*. http://www.un.org/en/documents/udhr/. Accessed July 4, 2012.

Urban, Wayne J. 1989. "The Black Scholar and Intelligence Testing: The Case of Horace Mann Bond." *Journal of the History of the Behavioral Sciences* 25:323–34.

Van de Velde, Theo H. 1933. *Ideal Marriage: Its Physiology and Technique*. New York: Convici-Friedi, Inc.

Voeller, Bruce. 1990. "Some Uses and Abuses of the Kinsey Scale." In *Homosexuality/Heterosexuality: Concepts of Sexual Orientation*, ed. David P. McWhirter, Stephanie A. Sanders, and June Machover Reinisch, 32–38. New York: Oxford University Press.

Wake, Naoko. 2011. *Private Practices: Harry Stack Sullivan, the Science of Homosexuality and American Liberalism*. New Brunswick, N.J.: Rutgers University Press.

Walker, E. L. 1941. "The Terman-Miles 'M-F' test and the Prison Classification System." *The Pedagogical Seminary and Journal of Genetic Psychology* 59:27–40.

Walker, Francis Amasa. 1890. "Statistics of the Colored Race in the United States." *Journal of the American Statistical Association* 2:91–106.

Walker, Helen A. 1945. "The Role of the American Statistical Association." *Journal of the American Statistical Association* 40:1–10.

Wallin, Paul. 1949. "An Appraisal of Some Methodological Aspects of the Kinsey Report." *American Sociological Review* 14:197–210.

Wallis, W. Allen. 1949. "Statistics of the Kinsey Report." *Journal of the American Statistical Association* 44:463–84.

Walsh-Bowers, Richard. 1999. "Fundamentalism in Psychological Science: The Publication Manual as 'Bible.'" *Psychology of Women Quarterly* 23:375–92.

Warner, Michael. 1993. "Introduction." In *Fear of a Queer Planet: Queer Politics and Social Theory*, ed. Michael Warner, vii–xxxi. Minneapolis: University of Minnesota Press.

Weeks, Jeffrey. 1977. *Coming Out: Homosexual Politics in Britain from the Nineteenth Century to the Present*. New York: Quartet Books.

Weidensall, Clara Jean. 1916. *The Mentality of the Criminal Woman*. Baltimore: Warwick & York.

Weinrich, James D. 1990. "The Kinsey Scale in Biology, with a Note on Kinsey as a Biologist." In *Homosexuality/Heterosexuality: Concepts of Sexual Orientation*, ed. David P. McWhirter, Stephanie A. Sanders, and June Machover Reinisch, 115–37. New York: Oxford University Press.

Wheeler, William Morton. 1920. "The Termitodoxa, or Biology and Society." *The Scientific Monthly* 10:113–24.

White, Hayden. 1973. *Metahistory: The Historical Imagination in Nineteenth-Century Europe*. Baltimore: Johns Hopkins University Press.

———. 1987. *The Content of the Form: Narrative Discourse and Historical Representation*. Baltimore: Johns Hopkins University Press.

White, John. 2006. *Intelligence, Destiny and Education: The Ideological Roots of Intelligence Testing*. London: Routledge.

White, Sheldon H. 1992. "G. Stanley Hall: From Philosophy to Developmental Psychology." *Developmental Psychology* 28:25–34.

Wiederman, Michael W. 1999. "Volunteer Bias in Sexuality Research Using College Student Participants." *Journal of Sex Research* 36:59–66.

Winston, Andrew. 1996. "'As His Name Indicates': R. S. Woodworth's Letters of Reference and Employment for Jewish Psychologists in the 1930s." *Journal of the History of Behavioral Sciences* 32:30–43.

———. 1998. "'The Defects of His Race:' E. G. Boring and Anti-Semitism in American Psychology, 1923–1953." *History of Psychology* 1:27–51.

Wissler, C. 1901. "The Correlation of Mental and Physical Tests." *Psychological Review: Monograph Supplements* 3(6):1–62.

Wood, James M., M. Teresa Nezworski, Scott O. Lillienfeld, and Howard N. Garb. 2003. *What's Wrong with the Rorschach? Science Confronts the Controversial Inkblot Test*. San Francisco: Jossey-Bass.

Woolgar, Steve, and Dorothy Pawluch. 1985. "Ontological Gerrymandering: The Anatomy of Social Problems Explanations." *Social Problems* 32:214–27.

Wright, Carroll D. 1990. *The History and Growth of the United States Census.* Washington, D.C.: Government Printing Office.

Yerkes, Robert Mearns. 1932. "Autobiography of Robert Mearns Yerkes." In *History of Psychology in Autobiography*, ed. C. Murchison, 2:381–407. Worchester, Mass.: Clark University Press.

Yerushalmi, Y. H. 1993. *Freud's Moses: Judaism Terminable and Interminable.* New Haven: Yale University Press.

Youden, W. J. 1951. "The Fisherian Revolution in Methods of Experimentation." *Journal of the American Statistical Association* 46:47–50.

Zborowski, Mark, and Elizabeth Herzog. 1952. *Life is with People: The Culture of the Shtetl.* New York: Schocken Books.

Zenderland, Leila. 1998. *Measuring Minds: Henry Herbert Goddard and the Origins of American Intelligence Testing.* Cambridge: Cambridge University Press.

Zimmerman, T. S., K. E. Holm, and S. A. Haddock. 2001. "A Decade of Advice for Women and Men in the Best-Selling Self-Help Literature." *Family Relations* 50:122–33.

Index of Names

General Index

143–49; findings of, 2, 25–27; humanistic
reactions to, 120; index of, 117–18; public
engagement with, 1, 68, 92, 98–100,
104–5, 111, 131, 148–49; sampling in, 47,
93, 101, 106, 119, 127, 132, 137–49, 180n37;
social scientists' reviews of, 140, 143–44.
See also Terman, Lewis: review of Kinsey's
SBHM in the Index of Names
sexuality: gender differences in, 26, 41–42,
93; historical change in practices, 46–49;
history of concepts, 11–12, 14, 22. *See also*
bisexual people; heterosexuality; homosexuality; masturbation; menstruation;
modernization: sexuality and; nocturnal
emissions; precocious development:
sexual
sexual outlet, 2, 10, 118. *See also* orgasm
sex with animals, 2
Shoah/Holocaust, 121
Sigma Xi, 127
slavery, 133
social class, 11, 14, 25–26, 35–36, 43, 73, 92,
96–97, 102, 126, 136, 154
social class mobility, 97, 104, 176n22. *See also*
social levels
social hygiene movement, 37–38, 52
social levels, 45–47. *See also* social class
Social Science Research Council, 141
Social Security Act of 1935, 135
sociology, 54
sperm, 28, 30–31, 153
Stanford University, 1, 5, 6–10, 22–23, 49–50,
53, 61–63, 78, 83, 86, 107–11, 141, 151,
167n8

statistics: education in, 102, 141; history of, 15,
131–36; inferential, 134–36, 138, 146, 158,
171n37; psychology of, 98–100; techniques, 81
stereotype threat, 162n15
sterilization, 50–51, 170n23
stupidity, 35–36, 52–53, 156
sublimation, 84, 89, 114, 126, 159–60

Talmud, 117, 118. *See also* Bibles
taxonomy: in animal biology, 70–72, 74–76,
173n14; in human sex research, 71–72,
98, 120, 138–39, 141, 146; "lumping" and
"splitting" in, 70–71, 75, 92, 144
Temple of the People, 87, 115
Terman's Kids (Shurkin), 64
testosterone, 153–54
transgender people, 82
trophylaxis, 72
tuberculosis, 33, 168n32

University of California, Berkeley, 86
University of Chicago, 143
University of Connecticut, 54

war, 2, 3, 38, 48, 72, 77, 81, 156
welfare, 154–55
West, North American, 20, 70–77, 119, 133–34
Western culture, 9, 13–14, 49, 114–15, 117,
123–24, 157, 160
White people, 1, 19, 31–32, 36–37, 38, 40,
51–52, 60, 92, 138, 140, 154, 34n35, 121n11,
139n35